Why & How Humans Created God

Fear, Death, Culture, and the Birth of the Divine

(Book 1)

Dr Prabhat Das

MD, PhD, FACP

No God Required

How Science Explains Reality

Without the Divine

Hardcover ISBN: 979-8-951067-00-5

Paperback ISBN: 978-1-971672-99-1

eBook ISBN: 978-1-971672-98-4

Printed in the United States of America

TABLE OF CONTENTS:

Author's Note

Why I Wrote This Series

Some books begin in libraries. Some begin in classrooms. This series began much earlier, in the mind of a curious boy growing up in India.

I was born into a Hindu religious family. Religion was not a distant theory in my childhood. It was part of daily life. It was in prayers, rituals, festivals, stories, customs, family conversations, and the quiet assumptions that surrounded me. God was not discussed as a possibility. God was treated as reality.

My father was a firm believer in an omnipotent God. To him, God was real, powerful, protective, and deeply involved in human life. His belief was sincere. It was part of his strength, his discipline, and his view of the world. I grew up respecting that seriousness.

But from my teenage years, another force began to grow inside me. Curiosity.

I wanted to know why things happened. I wanted to understand nature, disease, death, the stars, the human body, and the mind. I was not satisfied with an answer simply because it was old, sacred, or repeated by many people. I wanted to know whether it was true.

In high school, I became deeply impressed by the lives and writings of great scientists. I still remember going to the local library and borrowing books written by, or about, Newton, Darwin, Einstein, and others. Those books were not popular among students at that time. In fact, I may have been the first student to borrow some of them. They were covered with dust. Some had even been damaged by insects.

But to me, they were treasures. Those dusty books opened a new world.

Newton showed me that nature follows laws.

Darwin showed me that life has a history far deeper than ancient stories.

Einstein showed me that reality is stranger, deeper, and more beautiful than ordinary imagination.

These thinkers did not ask me to believe blindly. They invited me to understand. They did not frighten me with punishment. They opened the door to wonder. They showed me that truth does not become weaker when questioned. It becomes stronger.

That invitation changed my life.

My curiosity did not end with school. It continued through medical college, scientific training, clinical practice, and decades of work as a physician. In medicine, I learned the value of evidence. A diagnosis cannot be accepted because it feels comforting. A treatment cannot be trusted because it is traditional. A belief about the body must face reality. It must face observation, testing, correction, and outcome.

That same habit of mind gradually shaped the way I looked at religion & God.

Even now, at retirement age, I continue to read almost every day. Science, philosophy, history, religion, psychology, evolution, cosmology, neuroscience, and the human mind still fascinate me. The more I read, the more I see how far human knowledge has come. Many things once explained by God are now understood through natural causes. Thunder, disease, eclipse, famine, fertility, mental illness, death, life, and the vast universe itself no longer need supernatural explanation.

This book series, "**No God Required: How Science Explains Reality Without the Divine", is a 7 book series**, and is my attempt to share that long journey in clear and simple language.

I do not write these books to insult believers. Many believers are kind, sincere, moral, generous, and deeply thoughtful people. I know this from my own family, my own culture, my patients, my friends, and my life experience. Faith often gives people comfort, hope, discipline, community, and emotional strength.

But faith does not decide truth.

A belief may comfort us and still need examination. A tradition may be ancient and still be mistaken. A prayer may bring peace and still not prove that anyone is listening. A sacred story may be beautiful and still be human-made.

My goal is to understand how belief began, why it became powerful, and why it still feels so real to so many people.

Why did human beings create gods?

Why did religion become one of the strongest forces in history?

Why do beliefs learned in childhood feel like truth?

Why do different cultures create different gods?

Why did science gradually make divine explanations unnecessary?

Can human beings live meaningful, moral, compassionate, and courageous lives without God?

These are serious questions. They deserve serious answers.

This series is based on knowledge gathered from many great books, scholars, scientists, historians, philosophers, psychologists, and thinkers. I have tried to bring their insights together in language that an ordinary educated reader can understand. My hope is that these books will encourage readers to think more rationally, question inherited beliefs, respect evidence, and study these subjects more deeply.

I also hope they will show that life without God does not have to be empty, cold, selfish, or immoral.

A life guided by reason, compassion, truth, science, justice, and human responsibility can be deeply meaningful. We do not need divine command to care for one another. We do not need fear of hell to be moral. We do not need supernatural stories to feel wonder before the universe.

Human beings are already capable of love, courage, kindness, beauty, and truth.

The universe is already astonishing.

Life is already precious.

Our responsibility to one another is already urgent.

No God is required for that.

Introduction

Before Human Beings Had Science, They Had Fear

Imagine a small group of early humans standing under a violent night sky.

There is no science. No medicine. No astronomy. No microscope. No telescope. No germ theory. No knowledge of electricity, weather systems, infection, genetics, earthquakes, eclipses, or the brain.

Thunder shakes the ground. Lightning tears open the darkness. A child burns with fever. A woman dies in childbirth. The river dries. The crops fail. Animals vanish from the hunting ground. A father dies, and a few nights later, his grieving son sees him in a dream, speaking as if death had not taken him.

What could those people think?

They had no hospital, no blood test, no vaccine, no antibiotic, no weather forecast, no explanation for the shaking earth, the darkened sun, the sudden plague, or the dead who returned in sleep.

But they had fear.

They had memory.

They had imagination.

They had grief.

They had questions that would not leave them alone.

Why did this child die?

Why did rain come last year and not this year?

Why did lightning strike that tree?

Why did the dead return in dreams?

Why did pain come to good people?

Why did the sun disappear in the middle of the day?

Why are we here?

Why must we die?

The human mind cannot live easily inside unanswered terror. It searches for a cause. It looks for intention. It wants a hidden hand behind the event. When the world is frightening and no natural explanation is available, supernatural explanation becomes almost irresistible.

Out of fear, gods were born.

Out of death, souls were imagined.

Out of dreams, spirits appeared.

Out of thunder, divine anger was heard.

Out of disease, punishment and demons were created.

Out of helplessness, ritual became control.

Out of suffering, prayer became necessary.

And out of the deep human hunger for meaning, God became humanity's first great answer.

This book begins there.

Not with insult. Not with mockery. Not with easy anger. Religion is too important for that. Belief in God is one of the most powerful ideas human beings have ever created. It shaped civilizations, laws, empires, families, wars, moral codes, temples, music, poetry, guilt, sacrifice, hope, and death.

It comforted the grieving. It frightened the disobedient. It blessed kings. It guided tribes. It gave the poor hope and gave rulers authority. It told human beings where they came from, why they suffered, how they should live, and what might happen after death.

An idea this powerful deserves serious examination.

But serious examination does not mean surrender. Respect for human feeling does not require acceptance of supernatural claims. A belief can be beautiful and still be false. A story can comfort the heart and still fail as a description of reality. A tradition can be ancient and still be mistaken. A prayer can bring peace and still not prove that anyone is listening.

This is the central spirit of this book.

We will not begin by laughing at religion. We will begin by understanding it. Then we will ask the harder question:

Does religion make better sense as a message from God, or as a creation of the human mind?

The Question That Changes Everything

The central question of this book is simple:

Why and how did human beings create God?

This question is not asked with hatred. It is asked with curiosity, seriousness, and intellectual honesty. If human beings created God, then religion is not a window into heaven. It is a mirror held up to human fear, hope, imagination, culture, and need.

That possibility changes everything.

If God were discovered in the same way gravity, oxygen, electricity, germs, or DNA were discovered, we would expect clear evidence. We would expect reliable methods. We would expect people from different cultures, after careful observation, to move toward the same conclusion.

That is what happens in science.

A physician in India, America, Japan, or Brazil may speak a different language, but pneumonia is still pneumonia. Blood pressure is still blood pressure. DNA is still DNA. Gravity does not change with culture. The structure of the atom does not depend on childhood teaching. The laws of motion do not depend on the religion of one's parents.

But religion does not behave that way.

A child born in one family may grow up believing in one God. A child born elsewhere may believe in many gods. Another may believe in karma, spirits, ancestors, divine energy, rebirth, or no personal God at all. Another may believe that salvation depends on one prophet, one scripture, one savior, one ritual, one church, one temple, or one path.

These beliefs are usually not discovered independently. They are received. They are inherited. They are taught before they are examined.

This is one of the most important facts about religion.

Most people do not study all religions carefully and then choose the one best supported by evidence. Most people absorb the religion of their family, community, language, and culture. A child first trusts. Only later does the child question. By then, belief may already feel like identity. It may be wrapped in songs, holidays, food, grandparents, prayer, fear, guilt, love, marriage, funeral rituals, and childhood memories.

This does not make believers dishonest. It makes them human.

But it should make all of us cautious. If belief is shaped so strongly by birthplace and childhood, then religious certainty is less impressive than it first appears. It may tell us more about human inheritance than divine revelation.

This book explores that possibility.

God Was the First Answer to the Unknown

Long before science, God explained almost everything.

God explained the sky, thunder, rain, drought, fertility, birth, disease, dreams, earthquakes, eclipses, victory, famine, morality, suffering, death, and the existence of the world itself.

When human knowledge was small, God became a large explanation.

This was understandable.

Imagine a mother watching her child die of fever thousands of years ago. She does not know about bacteria, viruses, dehydration, meningitis, pneumonia, sepsis, or the immune system. She only knows that the child was alive yesterday and is silent today.

The death feels personal. It feels like a message. It feels like punishment, curse, fate, divine will, or the work of an invisible force.

Now imagine a farmer watching the sky during a long drought. The soil cracks. The animals weaken. The children grow thin. No one knows about climate patterns, ocean temperatures, atmospheric circulation, soil conditions, or long-term weather cycles. The farmer prays because prayer feels like action when there is nothing else to do. The community sacrifices because sacrifice feels like negotiation with power.

Now imagine a man dreaming of his dead father. In the dream, the father is alive, speaking, walking, recognizable. The dream is vivid. It shakes the heart. No one knows about REM sleep, memory networks, grief, or the brain's power to build images during sleep. The simplest explanation may be that the dead are not fully dead. Perhaps something survives. Perhaps there is a soul. Perhaps there is another world.

In such a world, religion is not strange. It is expected.

The early human mind was not foolish. It was trying to survive. It was trying to explain danger. It was trying to reduce terror. It was trying to protect the tribe. It was trying to find meaning in pain. It was trying to make death bearable.

God was humanity's first great answer to fear, death, and mystery.

But the first answer is not always the true answer.

Comfort Is Powerful, but It Is Not Evidence

One reason religion survived is that it did not merely explain.

It comforted.

A universe without God can feel cold to many people. Death can feel final. Suffering can feel meaningless. Injustice can feel unbearable. Religion softens these fears. It says the dead are not truly gone. It says suffering will be rewarded. It says evil will be punished. It says someone is watching. It says someone cares. It says life has a plan.

These promises are emotionally powerful.

A grieving mother does not want a lecture on biology. She wants her child back. A dying man may not want a discussion of neuroscience. He wants hope. A poor laborer crushed by injustice may want to believe that some final court exists beyond this life, where every hidden cruelty will be exposed and every tear will be counted.

We must understand this. If we do not understand the emotional need behind religion, we will never understand why religion became so strong.

But emotional need cannot decide what is true.

A belief may reduce fear and still be false. A hope may comfort us and still be unsupported. A story may help a person endure pain and still not describe reality.

As a physician, I have seen how deeply human beings need comfort. When people are sick, frightened, or dying, they reach for meaning. That is natural. That is human.

But in medicine we also learn a hard lesson: comfort cannot replace diagnosis. Hope cannot replace evidence. A beautiful belief cannot cure infection, dissolve a blood clot, reverse sepsis, repair a damaged heart, or change the result of a biopsy.

Reality does not become true because we need it to be true.

That is one of the central lessons of science. It is also one of the central lessons of this book.

Religion Gave People More Than Answers

Religion did not survive only because it gave explanations. It survived because it served many human needs at the same time.

It explained nature.

It comforted grief.

It reduced fear.

It gave moral rules.

It created group identity.

It strengthened obedience.

It gave authority to leaders.

It gave rituals to the community.

It gave people a story larger than themselves.

That is why religion became so deeply woven into human life.

A person did not simply believe in God as an isolated idea. God came wrapped in family, festivals, songs, food, marriage, funerals, childhood memories, social approval, fear of punishment, hope of reward, and loyalty to ancestors. Religion became part of identity.

That is why questioning religion can be so painful.

When people question God, they may feel they are questioning their parents, their childhood, their community, their marriage, their moral worth, their dead relatives, and their place in the world. They may fear losing family. They may fear losing certainty. They may fear losing heaven. They may fear hell. They may fear becoming immoral. They may fear becoming empty.

This explains why religious disagreement can become so emotional. For many people, God is not only an idea. God is home.

This book will treat that reality with respect. But it will also ask whether a belief should be protected from examination simply because it is emotionally important.

The answer must be no.

The more important a belief is, the more carefully it should be examined.

Why Belief Feels Stronger Than Argument

Many believers say, "I know God is real because I feel His presence."

This feeling can be powerful. A person may feel God in prayer, music, worship, silence, nature, grief, danger, or sudden relief. The experience may bring tears, peace, strength, courage, or a sense of being loved.

But human beings can feel many things with great intensity.

A dream can feel real.

A memory can feel certain and still be inaccurate.

A coincidence can feel like a sign.

A ritual can create deep emotion.

Music can change mood.

A crowd can intensify belief.

Grief can make the dead feel near.

Fear can make an unseen protector feel necessary.

The human brain does not simply record reality like a camera. It interprets. It connects. It fills gaps. It creates stories. It gives meaning. It looks for intention. It turns random events into patterns.

This ability helped our ancestors survive. A mind that quickly detected danger and hidden causes was more likely to survive in a dangerous world. But the same ability can also mislead us. It can make us see purpose where there is no purpose. It can make us feel watched when no one is watching. It can make us hear messages in coincidence.

This does not mean religious experiences are fake in the sense of being deliberately invented. They may be sincere and emotionally real. But the question is not whether the experience feels real. The question is whether the experience proves an external God.

That is a much harder question.

This book will argue that the feeling of God can be explained by the human mind itself: by fear, grief, hope, memory, ritual, culture, expectation, and social reinforcement.

God may feel real because the human mind is powerful enough to create the experience of God.

The Hidden Power of Childhood Belief

Most people do not choose their religion after careful examination of all religions.

Most people inherit religion.

A child born into a Hindu family is more likely to become Hindu. A child born into a Muslim family is more likely to become Muslim. A child born into a Christian family is more likely to become Christian. A child born into a Jewish, Buddhist, Sikh, Jain, or secular family is usually shaped by that environment.

It is a fact of human culture.

Family is powerful. Language is powerful. Repetition is powerful. Fear is powerful. Love is powerful. Childhood memory is powerful. A belief placed into the mind early may feel like truth for the rest of life.

By the time questioning becomes possible, religion may already be tied to identity. It may be connected to the mother's prayer, the father's instruction, the grandmother's story, the festival meal, the funeral ritual, the sacred song, the first fear of sin, the first idea of heaven, and the first warning about hell.

That is why religious belief often feels self-evident to the believer. It was absorbed before it was analyzed.

One of the goals of this book is to help the reader step back and ask:

Did I discover this belief, or did I receive it?

Did I reason my way to it, or was I trained into it before I knew how to reason?

Would I hold the same belief if I had been born in another country, another family, another century?

These are not insulting questions. They are honest questions.

What This Book Does Not Claim

This book does not claim that every religious person is bad. That would be false.

It does not claim that religion has never done good. That would also be false.

It does not claim that science has already answered every question. It has not.

It does not claim that all religious stories are worthless. Many contain poetry, beauty, moral struggle, historical memory, and emotional truth.

It does not claim that human beings no longer need meaning, comfort, or moral seriousness. We need them deeply.

The claim is more precise:

Human beings created God because God answered powerful human needs. But human need is not evidence of divine reality.

This distinction matters.

A child may need protection. That does not prove that an invisible protector exists. A grieving person may need reunion. That does not prove an afterlife. A suffering person may need justice. That does not prove a divine judge. A frightened society may need moral order. That does not prove a cosmic lawgiver.

Need explains belief.

It does not prove truth.

The Place of This Book in the Series

This book is the first of seven volumes in the series "No God Required: How Science Explains Reality Without the Divine".

The full series will examine how science and reason have replaced divine explanation in many areas of life.

This **first book** asks why and how humans created God.

The **second book** will examine how science made God unnecessary as an explanation through major discoveries and the work of major scientists.

The **third book** will examine the universe without a creator.

The **fourth book** will examine life without a designer.

The **fifth book** will examine mind without soul and morality without God.

The **sixth book** will examine the failure of the personal God through evil, hiddenness, prayer, miracles, scripture, and the silence of heaven.

The **seventh book** will bring the full argument together in the scientific case for atheism.

The journey begins here, with the human origin of God.

Before we can understand why God is no longer required, we must understand why God was created in the first place.

How to Read This Book

This book should be read with an open but critical mind.

If you are already an atheist, this book may help you understand religion more deeply and more humanely.

If you are agnostic, this book may help you see why belief in God is not necessary to explain religious experience.

If you are a believer, this book may challenge you, but it is not written to insult you. It is written to ask serious questions about the origin of belief.

If you are a doubter, this book may give language to thoughts you have already had but were afraid to say clearly.

The book asks you to consider one powerful possibility:

Perhaps God was not the source of human religion.

Perhaps human religion was the source of God.

That possibility is not small. It changes everything.

Transition to Chapter 1

We begin where humanity began: in a world without science.

Imagine again that ancient human standing under a violent sky. Thunder shakes the ground. Lightning splits the darkness. A child is sick. The crops are failing. The dead appear in dreams. No one knows about electricity, microbes, weather systems, brain activity, genetics, or astronomy.

In that world, gods were not strange.

They were almost inevitable.

God did not enter the human mind first as evidence.

God entered as rescue.

To understand God, we must first understand the world that made God necessary.

Sources

Atran, Scott. In Gods We Trust: The Evolutionary Landscape of Religion. Oxford University Press, 2002.

Boyer, Pascal. Religion Explained: The Evolutionary Origins of Religious Thought. Basic Books, 2001.

Dennett, Daniel C. Breaking the Spell: Religion as a Natural Phenomenon. Viking, 2006.

Hume, David. The Natural History of Religion. 1757.

Shermer, Michael. The Believing Brain: From Ghosts and Gods to Politics and Conspiracies. Times Books, 2011.

Chapter 1: Before Science, There Were Gods

The Night Before Knowledge

Imagine a small human group gathered near a fire on a cold, dangerous night.

The fire is the only circle of safety. Beyond it, the darkness is alive with sound. Leaves move. Animals cry. A branch breaks somewhere outside the light. The wind carries the smell of wet earth. Above them, the sky is turning black. A storm is coming.

Then lightning cuts across the clouds.

For one bright second, the whole world appears: trees, rocks, frightened faces, shining eyes. Then darkness returns. Thunder follows, deep and violent, rolling across the land like a voice too large for human beings.

Inside the shelter, a child is burning with fever.

His mother holds him close. His skin is hot. His breathing is fast. His small body trembles. No one knows what fever is. No one knows about bacteria, viruses, parasites, inflammation, dehydration, pneumonia, malaria, meningitis, or sepsis. No one knows that invisible organisms can enter the body and multiply. No one knows that the body has an immune system fighting a hidden battle inside the blood.

The mother knows only this: yesterday her child was laughing, and tonight he may die.

Outside, thunder shakes the ground again.

Someone says the sky is angry. Someone says the ancestors are warning them. Someone says an offering must be made. Someone says a rule must have been broken. Someone says the child has been touched by a spirit.

The mother prays, not because she has evidence that someone is listening, but because prayer feels like the only action left when knowledge has ended.

This is where religion begins.

Not in a university. Not in a calm philosophical debate. Not in a laboratory. Religion begins in fear, darkness, grief, helplessness, hunger, danger, and the urgent human need to explain what cannot yet be understood.

Before science, the world did not feel neutral. It felt personal. The storm seemed to speak. The river seemed to give and take. Disease seemed to punish. Dreams seemed to open a door. The dead seemed not fully gone. Nature seemed filled with intention.

Human beings did not create gods because they were foolish. They created gods because they were human. They were frightened, intelligent, imaginative, and desperate for meaning in a world that gave them no clear answers.

This chapter begins with that world.

A World Without the Tools of Truth

To understand why gods appeared in the human imagination, we must first remove almost everything modern knowledge has given us.

Remove the microscope.

Remove the telescope.

Remove the hospital.

Remove antibiotics, vaccines, anesthesia, surgery, blood tests, X-rays, CT scans, public health, genetics, weather satellites, geology, astronomy, and modern psychology.

Remove the knowledge that the Earth moves around the Sun. Remove the knowledge that lightning is electrical discharge. Remove the knowledge that infectious diseases can be caused by microbes. Remove the knowledge that earthquakes come from movements of the Earth's crust. Remove the knowledge that dreams are created by the sleeping brain. Remove the knowledge that eclipses are predictable alignments of the Sun, Moon, and Earth.

Now imagine trying to live.

A drought lasts for months. The ground cracks. Animals weaken. Children cry from hunger. The people ask, "Why has the sky abandoned us?"

A woman dies during childbirth. The family asks, "What power took her?"

A hunter is killed by an animal. The tribe asks, "What sign did we miss?"

A sickness spreads through the village. One house loses three people. Another house is spared. The people ask, "Why them and not us?"

The Sun suddenly darkens in the middle of the day. Birds go quiet. The air changes. People look upward in terror. They ask, "Who is attacking the Sun?"

These questions demanded answers. Human beings cannot easily live inside raw uncertainty. The mind reaches for cause. The heart reaches for comfort. The group reaches for ritual. When natural causes are unknown, supernatural causes become emotionally attractive.

This does not prove that gods exist.

It explains why gods were imagined.

Religion was not born in a world of evidence. It was born in a world of mystery.

Nature Did Not Feel Silent

To the modern mind, nature is powerful, but not personal. A hurricane does not hate us. A virus does not punish us. A volcano does not choose its victims. A drought does not arrive because a child disobeyed his parents.

But this way of thinking is recent.

For most of human history, nature felt alive. The sky could be angry. The sea could be hungry. The forest could be haunted. The mountain could be sacred. The river could be generous. Disease could be a curse. The harvest could be a blessing.

This was not irrational in the way we may imagine from our modern comfort. It was a natural attempt to understand a world filled with forces that moved, roared, killed, nourished, and disappeared without warning.

A storm has sound and movement. It comes from above. It flashes with light. It breaks trees, floods homes, and kills animals. To a prescientific mind, it can easily feel like action. And where there is action, the human mind often imagines an actor.

Thunder becomes a voice.

Lightning becomes a weapon.

Drought becomes displeasure.

Flood becomes punishment.

Fertility becomes blessing.

Disease becomes curse.

The dead appearing in dreams become spirits.

This is one of the oldest habits of the human mind: to see agency behind events. We still do this in small ways. We say the weather is cruel. We say fortune smiled. We say illness attacked. We say fate was kind. For us, these are often figures of speech. For early human beings, they may have felt much closer to reality.

The world seemed full of hidden minds.

Gods, spirits, demons, ancestors, and unseen powers gave those hidden minds names.

Fear Needed a Face

Fear without a face is difficult to bear.

When danger has no explanation, it becomes even more frightening. A visible enemy can be watched. A known disease can be treated. A predictable storm can be prepared for. But an unknown force creates helplessness.

Gods gave fear a face.

If drought came from an angry rain god, perhaps the people could offer sacrifice. If disease came from a demon, perhaps a healer could drive it away. If the dead continued as spirits, perhaps they could be honored, fed, feared, or asked for help. If the tribe lost a battle because a god was displeased, perhaps obedience could restore favor.

A supernatural explanation gave people something to do.

They could pray.

They could sacrifice.

They could confess.

They could purify.

They could obey.

They could repeat sacred words.

They could follow a priest, shaman, elder, or prophet who claimed to understand the invisible world.

This was psychologically powerful. Even when ritual did not change nature, it changed the human feeling of helplessness. It gave structure to fear. It gave action to anxiety. It gave the group a shared response.

A child is sick, so people gather and pray.

The crops fail, so people perform a rain ritual.

A leader dies, so people bury him with objects for another world.

A plague spreads, so people search for sin, impurity, or divine anger.

From a scientific point of view, these actions do not explain disease, weather, or death. But from a human point of view, they reduce chaos. They make suffering feel connected to a larger story.

This is one reason religion became so durable. It did not only offer answers.

It offered action.

The First Explanations Had Human Faces

Early explanations of nature often used human models.

Human beings understood intention because they had intentions. They understood anger because they felt anger. They understood hunger, jealousy, love, revenge, punishment, protection, command, and obedience because these were part of social life.

So when they tried to explain nature, they naturally used the language of human life.

The gods wanted.

The gods commanded.

The gods punished.

The gods loved.

The gods became angry.

The gods demanded offerings.

The gods chose one people over another.

The gods listened.

The gods judged.

The gods forgave.

In this way, the universe became a larger version of the tribe. There were rulers above, servants below, laws, rewards, punishments, loyalty, betrayal, anger, protection, and favor.

This was a deeply human way of explaining reality.

A river that floods is not merely water following gravity and rainfall. It becomes a force that gives life and takes life. A disease is not merely biological failure. It becomes punishment or possession. A successful harvest is not merely soil, weather, seed, and labor. It becomes divine favor.

The unknown world was interpreted through the known world of human relationships.

This is why gods so often resemble the societies that worship them. Warrior cultures imagine warrior gods. Agricultural societies imagine fertility gods. Hierarchical societies imagine kingly gods. Patriarchal societies often imagine male gods. Communities threatened by enemies may imagine protective gods. Societies built around law may imagine a divine judge.

Human beings made gods in the shape of their fears, hopes, and social arrangements.

That is not a small observation. It is one of the strongest clues that religion comes from human imagination.

Randomness Was Too Painful

Randomness is emotionally difficult.

If lightning strikes one house and not another, the human mind asks why. If one child dies and another survives, the mind asks why. If one prayer seems answered and another is not, the mind asks why. We do not easily accept that life can be uneven without a hidden reason.

The mind prefers a story.

A story says the child died because the gods were displeased.

A story says the drought came because the people failed in ritual.

A story says the enemy won because the tribe lost divine favor.

A story says a dream came because the dead wanted to speak.

A story says suffering has a purpose.

A story is often easier to bear than chance.

But the fact that a story comforts us does not make it true. Sometimes events happen because of physical causes without moral intention. Sometimes disease strikes because of infection, genetics, environment, or chance. Sometimes crops fail because of weather. Sometimes earthquakes occur because of geological movement. Sometimes good people suffer without cosmic purpose.

This is one of the hardest truths for the human mind to accept.

Religion often protects people from that truth. It turns randomness into purpose. It turns accident into plan. It turns suffering into test. It turns death into passage. It turns uncertainty into destiny.

That is emotionally powerful.

But emotional power is not evidence.

A story may make pain easier to carry, but it may still be only a story.

Stories Became the First Textbooks

Before scientific method, stories became the containers of explanation.

A story could explain where the world came from.

A story could explain why humans die.

A story could explain why women suffer in childbirth.

A story could explain why snakes are feared.

A story could explain why the sea is dangerous.

A story could explain why the tribe is special.

A story could explain why a mountain is sacred.

A story could explain why one food is forbidden and another is holy.

A story could explain why a king has divine authority.

Stories are powerful because they are easier to remember than abstract principles. They can be told around fires, repeated to children, sung in rituals, carved into temples, and preserved through festivals. A good story does not need laboratory proof to survive. It needs emotional force, repetition, authority, and community.

This is why myths became so important.

A myth is not merely a false story. It is a story that gives meaning to a group. It tells people who they are, where they came from, what they owe the gods, why they suffer, and what happens after death.

Myths helped human beings live before science gave them better explanations.

But myths also created a problem. Once a story becomes sacred, it becomes difficult to question. Once a story is tied to ancestors, identity, law, and fear, doubt begins to feel like betrayal. The child who questions the story may be accused of questioning the tribe itself.

This is how early explanations became protected from correction.

Science would later create a different habit: do not protect the story.

Test it.

The Sacred Grew From Human Meaning

Not every rock is sacred. Not every tree is sacred. Not every hill is sacred. Yet in many cultures, certain places become charged with special meaning.

A cave where ancestors were buried.

A river that gave life.

A mountain where storms gathered.

A grove where rituals were performed.

A stone that marked a death.

A place where a vision was reported.

A battlefield where many died.

Human beings do not live in physical space only. They live in meaningful space. We attach memory to places. We attach grief to objects. We attach hope to rituals. We mark birth, marriage, and death with symbols because symbols help us carry emotion.

The sacred may have begun this way: ordinary parts of nature became filled with human meaning.

Over time, meaning became mystery. Mystery became reverence. Reverence became ritual. Ritual became tradition. Tradition became religion.

This does not mean sacred places have no emotional value. They clearly do. A grave matters. A home matters. A place of childhood matters. A national monument matters. A temple, mosque, church, shrine, or sacred river may matter deeply to believers.

But emotional value is not the same as supernatural proof.

A place can be sacred to human memory without being touched by a god.

This distinction will appear again and again in this book. Human beings can create real meaning. That does not mean they have discovered supernatural reality.

Gods Multiplied Because Human Lives Differed

If religion came from one clear divine source, we might expect one clear religion.

But human history shows the opposite.

Gods multiplied. Spirits multiplied. Myths multiplied. Rituals multiplied. Sacred laws multiplied. Afterlife stories multiplied. Creation stories multiplied.

There were gods of the sky, gods of the sea, gods of war, gods of fertility, gods of death, gods of harvest, gods of love, gods of rivers, gods of mountains, gods of fire, gods of ancestors, gods of cities, gods of kings, and gods of the underworld.

This variety makes sense if gods are created by human cultures. Different societies face different fears and needs. A desert people may long for water. A farming people may depend on fertility. A seafaring people may fear storms. A warrior society may honor strength. A kingdom may imagine heaven as a royal court. A tribe surrounded by enemies may imagine a protective divine warrior.

Religion reflects life.

This does not prove by itself that no god exists, but it strongly suggests that specific gods are shaped by human conditions. The gods of history look like cultural products. They carry the fingerprints of the societies that created them.

This is why the study of religion is also the study of humanity.

When we examine gods, we are often examining ourselves: our fears, our hopes, our politics, our grief, our need for control, and our longing for meaning.

Useful Does Not Mean True

Religion did not have to be true in order to be useful.

A belief can help a group survive even if the belief itself is false. A ritual can reduce anxiety even if it does not change the weather. A shared myth can unite a tribe even if the story never happened. A belief in watching gods can encourage obedience even if no god is watching. A promise of afterlife can reduce fear of death even if death is final.

Usefulness and truth are not the same.

This distinction is essential.

Religion may have helped early people cooperate, endure suffering, bury the dead, raise children, maintain social order, explain misfortune, and face death. These functions are real. They should not be dismissed.

But the fact that religion served human needs does not prove that its supernatural claims are true.

A medicine can comfort a patient and still be ineffective. A rumor can unite a group and still be false. A superstition can reduce anxiety and still have no effect on reality.

Science asks a different question. Not simply, "Does this belief comfort us?" Not simply, "Does this belief unite us?" Not simply, "Did our ancestors believe it?" Science asks, "What is the evidence? Can it be tested? Does it predict? Does it explain better than alternatives? Does it remain true when examined?"

By that standard, many ancient God explanations became weaker as knowledge grew.

The thunder was not a voice.

The eclipse was not a dragon.

The plague was not a curse.

The earthquake was not divine anger.

The dream was not proof of another world.

The fever was not a demon.

Again and again, natural explanations replaced supernatural ones.

But in the beginning, before science, gods filled the darkness.

Chapter Summary

Before science, human beings lived in a frightening and mysterious world. They did not know the natural causes of disease, storms, drought, earthquakes, eclipses, dreams, fertility, or death. The human mind searched for causes, patterns, and hidden intentions. In that world, gods, spirits, ancestors, demons, and sacred forces became emotionally natural explanations.

Religion began not because early humans were foolish, but because they were trying to survive uncertainty. Gods gave fear a face, suffering a story, ritual a purpose, and death a possible answer. Religion was useful because it comforted, organized, explained, and united human groups.

But usefulness is not truth. A belief can reduce fear and still be false. A story can comfort a tribe and still fail as an explanation of reality. This chapter has shown why gods were likely to appear in a world without science.

God was not born from evidence.

God was born where fear met imagination.

Key Points

Before science, natural events were often interpreted as supernatural actions.

The human mind searches for causes, patterns, and agency, especially when afraid.

Gods gave early humans explanations for disease, storms, death, dreams, fertility, and disaster.

Religion helped reduce helplessness by offering ritual, prayer, sacrifice, and shared stories.

Different cultures created different gods because they faced different fears, environments, and social needs.

Religion could be useful without being true.

God was humanity's early explanation for a world it did not yet understand.

Transition to the Next Chapter

The first and deepest mystery was not thunder, lightning, or eclipse.

It was death.

Human beings do not merely die. They know they will die. They remember the dead. They dream of them. They bury them. They speak their names. They refuse, emotionally, to let them disappear completely.

From this pain came one of religion's most powerful promises: death is not the end.

The next chapter examines how fear of death helped create the dream of immortality.

Sources

Atran, Scott. In Gods We Trust: The Evolutionary Landscape of Religion. Oxford University Press, 2002.

Boyer, Pascal. Religion Explained: The Evolutionary Origins of Religious Thought. Basic Books, 2001.

Hume, David. The Natural History of Religion. 1757.

Tylor, Edward B. Primitive Culture. 1871.

Frazer, James George. The Golden Bough: A Study in Magic and Religion. 1890.

Chapter 2: Fear of Death and the Dream of Immortality

The Only Animal That Knows It Will Die

Every living thing dies.

Leaves fall. Flowers dry. Insects vanish. Birds drop from the sky. Fish float to the surface. Deer are taken by wolves. Old trees collapse. Even stars burn out after long ages of fire.

But human beings carry a special burden.

We do not merely die.

We know we will die.

That knowledge changes everything.

A deer may run from a predator. A bird may avoid danger. A dog may grieve when its companion is gone. Many animals show fear, attachment, and loss. But human beings do something more painful. We can sit in silence and imagine our own absence. We can stand beside a dead body and know that one day others may stand beside ours. We can look at an aging face in the mirror and understand that time is carrying us toward an ending we cannot refuse.

This awareness is one of the deepest roots of religion.

Human beings love. They remember. They plan. They dream. They forgive. They regret. They carry unfinished conversations inside the heart. It feels almost impossible to believe that all of this can simply stop. The body may die, but the mind protests. Love refuses to accept silence. Grief searches for a door.

Religion opened that door.

It said the dead are not truly gone.

It said the body dies, but the soul survives.

It said the grave is not the end.

It said the mother, father, child, spouse, friend, and self continue somewhere beyond sight.

This promise became one of religion's greatest powers.

Without death, religion would still exist in some form. Human beings would still fear storms, disease, hunger, disaster, injustice, and uncertainty. But religion would be far weaker without death. Death gives religion its deepest urgency. It gives the priest, prophet, scripture, ritual, temple, and prayer their most powerful role.

Religion does not only explain the world.

It promises that we do not disappear.

That promise can comfort the heart. But comfort is not proof. The central question remains: did human beings discover immortality, or did fear of death create the dream of immortality?

The Body Was Still, but Love Was Not

Imagine an ancient family gathered around a dead father.

His body is still there. His face is familiar. His hands are the same hands that once carried wood, lifted food, held a child, built shelter, touched a shoulder, and pointed toward danger. But now those hands do not move. The voice is gone. The eyes do not open. The body that once answered now gives no answer.

For the living, this creates a terrible puzzle.

The body is present, but the person seems absent.

Where did he go?

That question may be one of the oldest religious questions.

To modern science, the answer is emotionally painful but biologically clear. The person we knew depended on the living brain. Memory, language, personality, desire, recognition, movement, emotion, and awareness required a functioning nervous system. When the brain permanently stops, the conscious person is gone.

But early humans did not know this.

They saw the body, but they could not see the brain's inner work. They did not know how memory is formed, how speech is produced, how personality depends on neural circuits, or how consciousness fades when the brain is injured, starved of oxygen, infected, poisoned, sedated, or destroyed.

They only knew that something had left.

The warmth had gone.

The breath had stopped.

The voice had disappeared.

The eyes no longer recognized.

The living presence was missing.

It was natural to imagine that an invisible part had departed. The breath-self, spirit, soul, or shadow had escaped the body. The body remained, but the person had gone elsewhere.

Many cultures connected life with breath. This is understandable. Breath is invisible, yet real. It is felt on the skin. It appears in cold air. A living person breathes. A dead person does not. Breath may have seemed like the invisible force of life itself.

From such observations, the idea of spirit could grow.

The body dies, but perhaps the breath-self continues.

The body decays, but perhaps the inner person survives.

The visible person disappears, but perhaps the invisible person remains.

This was not a scientific conclusion. It was a human answer to a wound.

It made death less final.

It allowed love to continue speaking.

It allowed grief to hope.

Dreams Made Death Look Temporary

Dreams may have played a major role in the birth of afterlife belief.

A person dies. The body is buried or burned. The family grieves. Days pass. Then one night, the dead person appears in a dream.

He speaks.

He walks.

He smiles.

He gives advice.

He asks for food.

He warns of danger.

He seems alive.

For us, with modern knowledge, a dream is an activity of the sleeping brain. It can be shaped by memory, emotion, grief, fear, and desire. But to early humans, dreams were powerful and mysterious. They did not have the language of REM sleep, neural activation, memory consolidation, or unconscious imagery. A dream was not easily dismissed as "only a dream." It was an experience.

If the dead appear in dreams, perhaps they still exist.

If a sleeping person travels in dreams while the body remains still, perhaps there is a second self that can leave the body.

If a dead mother appears and comforts her child, perhaps she is not fully gone.

If a dead father speaks, perhaps death has not ended the relationship.

This is how an invisible world becomes believable.

The dream world is strange. People who are dead can live there. Places change instantly. A person can fly, fall, speak with animals, meet old friends, and feel terror or joy while lying still in darkness.

For early humans, dreams may have made the boundary between visible and invisible reality feel thin. The sleeping body remained near the fire, but the dream self wandered. This could easily support the idea that human beings have an inner double, a spirit, a soul, or a self that can leave the body.

Dreams of the dead would have been especially powerful because grief already longs for reunion. The dream gives reunion. The heart wants the dead to survive. The dream seems to show survival. The mind wants death to be temporary. The dream makes death look temporary.

This does not mean early humans were foolish. They were interpreting experience with the tools they had. The emotion was real. The grief was real. The dream was real as an experience.

But the interpretation may have been wrong.

A dream can reveal love without revealing heaven.

A dream can show grief without proving survival after death.

A dream can feel like visitation and still arise from memory.

The human mind can create powerful inner worlds. Religion often mistakes those inner worlds for external reality.

The Soul Was Born From the Refusal to Vanish

The idea of the soul is one of the most beautiful and persistent ideas in human history.

It says that we are not merely bodies. It says that something inside us is deeper than flesh, stronger than decay, and more lasting than death. It says that the person we love is not destroyed when the heart stops. It says that the child, mother, father, friend, lover, and self continue in another form.

No wonder the idea became powerful.

The soul answers the unbearable question: what happens to me when my body dies?

It gives a comforting answer: you continue.

The body is temporary, but the soul is permanent.

The body weakens, but the soul remains.

The body enters the grave, but the soul travels on.

This idea appears in many forms across cultures. Some traditions imagine a spirit world. Some imagine ancestors watching the living. Some imagine heaven. Some imagine hell. Some imagine rebirth. Some imagine judgment. Some imagine union with a divine reality. Some imagine the soul returning again and again until it is purified or liberated.

The details differ, but the emotional center is similar:

Death does not erase us.

This is a powerful human hope.

But when examined carefully, the soul raises difficult questions.

If the soul carries the person, why does the person change when the brain changes? Why can a head injury alter personality? Why can dementia erase memory? Why can a tumor change behavior? Why can anesthesia switch off awareness? Why can drugs, fever, oxygen loss, stroke, infection, or trauma change the mind so deeply?

If the soul is the true self, why does the self depend so strongly on the brain?

This question will be examined more fully in a later book in this series, Mind Without Soul, Morality Without God. But it is important to introduce here because the idea of the soul likely arose long before human beings understood the brain.

Early humans saw life leave the body. They saw breath stop. They saw dreams of the dead. They felt that love could not simply end. From these experiences, the soul became believable.

But the soul may tell us more about the human fear of disappearance than about the actual survival of consciousness.

It may be one of humanity's most tender inventions.

Tender, but not proven.

Ancestors Who Refused to Leave

Many religions began not with a distant creator God, but with the dead.

The dead remained important. They were remembered, feared, honored, fed, buried with objects, invoked in rituals, and sometimes treated as protectors or judges.

Ancestor veneration is one of the oldest and most widespread forms of religious life.

This makes human sense.

The dead do not disappear emotionally when they disappear physically. A father may be dead, but his commands remain in memory. A mother may be gone, but her voice is still heard in the mind. A grandparent may be buried, but the stories continue. The dead remain inside the living through memory, habit, fear, gratitude, and love.

A family may ask: What would father think? What would mother want? What did our ancestors teach? What do we owe to those who came before?

Over time, memory can become presence.

Presence can become spirit.

Spirit can become worship.

Before one all-powerful God, there were often many dead relatives, many spirits, many ancestors, many invisible presences.

Burial practices also suggest that early humans took death seriously in ways that went beyond simple disposal of bodies. Graves, objects placed with the dead, ritual treatment of bodies, and signs of mourning suggest that death was surrounded by meaning. We must be careful in interpreting ancient burials because we cannot read the minds of people who lived thousands of years ago. But it is reasonable to say that death became one of the earliest centers of symbolic and religious behavior.

A dead body forced a question.

Is this the end?

The human heart often answered no.

Not because it had evidence, but because it had love.

Ancestor belief allowed the dead to remain socially alive. They continued to matter. They continued to belong. They continued to guide the living. In this way, religion gave human memory a supernatural frame.

It allowed families and tribes to say: we are not alone. Those who came before us are still near.

Heaven Was the Great Human Protest

Heaven is one of the most emotionally powerful ideas ever created.

It is a protest against death.

It says the grave does not win. It says the child who died is safe. It says the mother will be seen again. It says the good person who suffered will be rewarded. It says the injustice of this world will be corrected in another world.

For people crushed by grief, poverty, disease, oppression, or violence, this promise is enormous.

If life is unfair here, heaven promises fairness there.

If bodies break here, heaven promises restoration there.

If tyrants escape justice here, heaven promises judgment there.

If love is separated here, heaven promises reunion there.

This is why heaven has such deep emotional force. It does not simply describe another place. It heals wounds that life refuses to heal.

But again, we must separate emotional beauty from evidence.

The fact that heaven answers a human need does not prove that heaven exists. In fact, the perfect fit between human fear and religious promise may suggest the opposite: heaven may have been shaped by the human longing it satisfies.

A thirsty person imagines water.

A lonely person imagines companionship.

A grieving person imagines reunion.

A frightened person imagines protection.

A mortal person imagines immortality.

Heaven may be the dream that death forced the human mind to create.

This does not make the dream meaningless. It may have helped millions endure suffering. It may have softened grief. It may have inspired courage. It may have given hope to the powerless.

But hope is not the same as truth.

A promise should not be accepted only because we desperately need it.

Hell Turned Death Into Control

If heaven comforted, hell controlled.

The idea of punishment after death gave religion enormous power. It extended authority beyond the visible world. A human ruler can punish the body, but a divine ruler can punish the soul. A court can judge behavior, but God can judge thought, desire, doubt, and secret guilt.

This is a powerful instrument of control.

A person may escape the tribe, the priest, the king, the police, or the family, but cannot escape an all-seeing God. Even death does not end accountability. In some religious traditions, death becomes the beginning of punishment.

Hell answers one human desire: the desire for justice. People want evil to be punished. They want cruelty, murder, betrayal, and oppression to face consequences. In a world where many wrongdoers escape, hell offers final justice.

But hell also creates deep fear.

Children may be taught that disbelief, disobedience, wrong ritual, wrong doctrine, wrong desire, or belonging to the wrong religion can lead to eternal suffering. The fear can be severe. It can shape identity, sexuality, obedience, guilt, and mental life.

There is also a moral problem. Eternal punishment for a finite human life raises serious questions about justice and proportion. How can endless suffering be a fair response to limited human error? How can a loving God create or permit such a fate? How can belief under threat be called free love?

These questions will be examined more fully in Book 6, The Failure of the Personal God. But here, in the story of religion's origin, we can see how powerful the idea of hell became.

Heaven offered comfort.

Hell enforced obedience.

Together, they made death the doorway through which religion gained influence over life.

Death Made Religion Urgent

A religion that speaks about weather may be important.

A religion that speaks about morality may be influential.

But a religion that claims power over death becomes urgent.

This is why death-centered beliefs are so powerful. If religion can tell people what happens after death, then religion becomes more than culture. It becomes the guide to ultimate destiny.

What must I believe?

What must I do?

What must I avoid?

What ritual must be performed?

What priest must be trusted?

What scripture must be obeyed?

What sin must be confessed?

What God must be worshiped?

The fear of death makes these questions feel urgent.

If this life is brief and the next life is eternal, then religion becomes more important than food, career, politics, wealth, or pleasure. It becomes the central business of life.

This is one reason religion can hold such power over human beings. It claims authority over the one thing no one can avoid.

Death comes to everyone: rich and poor, king and laborer, scholar and child, believer and unbeliever. A belief system that promises to explain death and defeat it holds tremendous emotional power.

But we must ask a sober question.

Did religion discover life after death, or did fear of death create belief in life after death?

This book argues that the second explanation is more plausible.

The human fear of death is real. The grief is real. The longing is real. The comfort is real. But none of these proves that consciousness survives the death of the brain.

They prove something else: human beings are meaning-seeking animals who cannot easily accept disappearance.

A Brief Life Can Still Be Deep

To question life after death is painful.

It is easier to question thunder gods, river spirits, or ancient myths than to question the hope of seeing our loved ones again. The idea of final death touches the deepest nerve in the human heart.

No honest atheist should treat this lightly.

When people believe in heaven, they are often not defending an abstract theory. They are defending their mother, father, child, spouse, friend, and their own hope of survival. To challenge afterlife belief is to touch grief itself.

That is why this discussion must be handled with tenderness.

But tenderness does not require dishonesty.

If something is not supported by evidence, we should not call it true simply because it is painful to doubt. If the dead live only in memory, influence, and the continuing effects of their lives, then that truth must be faced. It may be difficult, but it may also be clarifying.

A life that ends is not meaningless.

A song is not meaningless because it ends.

A flower is not meaningless because it fades.

A friendship is not meaningless because time is short.

A sunset is not meaningless because darkness follows.

Meaning does not require eternity. Sometimes the fact that life is brief makes it more precious, not less.

Religion tried to defeat death by imagining another world. But there is another way to answer death: by living more honestly, loving more deeply, reducing suffering where we can, and recognizing that this life, the only life we know we have, deserves our full attention.

The dream of immortality may comfort us.

But courage begins when we ask whether the dream is true.

Chapter Summary

Death is one of the deepest roots of religion. Human beings are aware of their own mortality, and this awareness creates fear, grief, longing, and the desire for continuity. Early humans did not understand the brain, consciousness, dreams, or the biological finality of death. The stillness of the dead body, the disappearance of breath, dreams of lost loved ones, and the emotional refusal to let go all helped create belief in souls, spirits, ancestors, heaven, rebirth, and immortality.

Religion became powerful because it promised that death was not final. Heaven offered comfort. Ancestor belief preserved connection. The soul gave hope that the self could survive the body. Hell added moral control and fear. Together, these ideas gave religion enormous authority over human life.

But the emotional need for life after death does not prove that life after death exists. Human longing explains why the belief arose. It does not prove the belief true.

The grave frightened humanity.

Religion answered with a door.

Science asks whether that door is real.

Key Points

Human beings carry the unique burden of knowing they will die.

Fear of death helped create belief in souls, spirits, ancestors, heaven, rebirth, and immortality.

Dreams of the dead may have strongly encouraged belief in an invisible world.

The idea of the soul likely grew from the sense that something invisible leaves the body at death.

Ancestor belief helped keep the dead emotionally and socially present among the living.

Heaven comforted grief, while hell strengthened religious control.

Emotional comfort is not the same as evidence.

Meaning does not require immortality.

Transition to the Next Chapter

Death frightened the human heart.

Nature frightened the human body.

The sky roared. Lightning struck. Rivers flooded. Crops failed. Women died in childbirth. Plagues spread. The ground shook. The Sun disappeared during eclipses. Nature seemed alive, unpredictable, and sometimes cruel.

Before science explained these forces, human beings often saw them as divine actions, curses, blessings, omens, or punishments.

The next chapter examines how frightening nature helped create gods.

Sources

Becker, Ernest. The Denial of Death. Free Press, 1973.

Boyer, Pascal. Religion Explained: The Evolutionary Origins of Religious Thought. Basic Books, 2001.

Davies, Douglas J. A Brief History of Death. Blackwell, 2005.

Lifton, Robert Jay. The Broken Connection: On Death and the Continuity of Life. Simon & Schuster, 1979.

Solomon, Sheldon, Jeff Greenberg, and Tom Pyszczynski. The Worm at the Core: On the Role of Death in Life. Random House, 2015.

Chapter 3: Nature Was Frightening Before It Was Explained

When the Sky Seemed Angry

Before science explained nature, the world did not feel like a system of physical forces.

It felt alive.

The sky roared. The river rose. The ground shook. The forest hid danger. The sea swallowed boats. The wind tore roofs away. The Sun disappeared during an eclipse. Rain saved a village, and the absence of rain slowly starved it.

To early human beings, nature was not a quiet background. It was a power above them, around them, and sometimes against them. It fed them, frightened them, injured them, and killed them. It gave food one season and famine the next. It brought children into the world, then took them with fever. It made the crops grow, then destroyed them with insects, drought, flood, or fire.

Human beings looked at this frightening world and asked a natural question:

Who is doing this?

That question reveals something important. When people do not understand impersonal forces, they often imagine personal forces. A storm becomes anger. A drought becomes punishment. A flood becomes judgment. A good harvest becomes blessing. A plague becomes curse. A healthy child becomes divine favor.

The natural world became a moral world.

This was one of the great steps in the birth of religion. Nature was not only powerful. It was interpreted. It was given intention. It was given emotion. It was given personality. Human beings did not merely observe nature. They entered into a relationship with it.

They feared it.

They thanked it.

They begged it.

They sacrificed to it.

They tried to please it.

This is how gods entered the sky, wind, river, mountain, fire, forest, womb, and grave.

Before nature was explained, nature was worshiped.

Thunder Was the Voice Above the World

Few natural events are more dramatic than thunder and lightning.

Even today, with all our scientific knowledge, a violent storm can make us pause. The sky darkens. The air changes. Lightning opens the clouds with sudden force. Thunder follows like a command. Windows tremble. Trees bend. Animals grow restless. Children become afraid.

Now imagine that same storm thousands of years ago.

There is no understanding of electricity. No knowledge of charged particles, atmospheric instability, or electrical discharge. No weather radar. No forecast. No lightning rod. No safe explanation.

The sky flashes, and something burns.

The sky speaks, and the ground shakes.

The sky strikes, and a tree splits open.

What could early human beings think?

To a frightened mind, thunder does not sound like a random physical process. It sounds like a voice. Lightning does not look like electrical discharge. It looks like a weapon. The storm does not feel like atmospheric movement. It feels like anger.

It is no surprise that many cultures imagined storm gods, sky gods, thunder gods, and lightning gods. The sky was high, unreachable, dangerous, and necessary. It gave rain, but it also gave destruction. It was above human control, and anything above human control easily becomes sacred.

Thunder may have been one of nature's first sermons. It taught human beings their smallness. It made them feel watched, judged, threatened, or warned. It turned the sky into a place of power.

Later science would change this. Lightning would become electricity. Thunder would become sound produced by rapidly expanding heated air. Storms would become weather systems. The sky would remain magnificent, but it would no longer need divine anger to explain it.

But before that knowledge, the sky had a voice.

And human beings listened.

Drought Turned the Sky Into a Judge

Rain is not merely weather.

For early human beings, rain was survival.

Without rain, seeds failed. Grass dried. Animals weakened. Rivers shrank. Milk disappeared. Hunger spread. Children cried. Old people died first. The village became quiet. The future became thin.

A drought is slow terror.

Unlike a predator, it cannot be fought. Unlike an enemy tribe, it cannot be chased away. Unlike a broken tool, it cannot be repaired. People can watch the sky for days, weeks, and months, and see nothing. The empty sky becomes an accusation.

Why is the rain not coming?

This question easily becomes religious.

Did we offend the gods?

Did we fail in ritual?

Did someone break a taboo?

Did an enemy curse us?

Do we need sacrifice?

Should we fast, pray, dance, confess, purify, or obey?

Rain rituals are understandable because drought creates helplessness. Ritual gives people something to do when there is nothing practical left to do. It turns waiting into action. It turns fear into performance. It turns anxiety into community.

When rain finally comes, the ritual seems confirmed.

People remember the prayer and the rain that followed. They may forget the prayers that did not bring rain. This is one way belief strengthens itself. The mind notices the meaningful coincidence and builds a story around it.

Rain becomes mercy.

Drought becomes displeasure.

The sky becomes a moral judge.

Modern science gives a different picture. Rain depends on water vapor, temperature, air pressure, wind patterns, geography, ocean currents, and climate systems. A drought is not punishment. It is not moral communication. It is a natural event with physical causes.

But long before meteorology, human beings did not see climate systems.

They saw the sky withholding life.

And they begged it to forgive them.

Disease Was a Curse Before It Was a Diagnosis

Disease was one of the greatest engines of supernatural belief.

Few experiences are more frightening than illness without explanation. A healthy person becomes weak. A child develops fever. A wound becomes swollen. A cough spreads through a family. A village is struck by plague. Some die quickly. Some survive. Some never become sick at all.

Before germ theory, this seemed deeply mysterious.

Why this person?

Why this house?

Why this child?

Why did one recover while another died?

Without knowledge of bacteria, viruses, parasites, contaminated water, immunity, nutrition, sanitation, genetics, and public health, disease easily became a spiritual event. It could be explained as curse, demon, impurity, sin, divine punishment, witchcraft, evil eye, fate, or test.

This was not merely ignorance. It was a human attempt to give meaning to suffering. Random illness is hard to accept. If disease is punishment, then at least it has a reason. If disease is caused by a demon, perhaps the demon can be expelled. If disease comes from impurity, perhaps purification may help. If disease is sent by God, perhaps prayer, sacrifice, or repentance may matter.

Supernatural explanations made illness feel morally organized.

But disease is not morally organized.

As a physician, I know this deeply. Disease does not choose the cruel and spare the kind. Cancer does not check virtue. Pneumonia does not ask about moral character. A virus does not know whether the body it enters belongs to a saint or a criminal. A genetic disorder does not punish the newborn for sin. An infection follows biology, not justice.

This is one of the most important ways medicine weakened religion. Once disease could be explained naturally, the old spiritual explanations lost much of their power. Germ theory, sanitation, vaccination, antibiotics, anesthesia, surgery, pathology, and public health transformed disease from curse into diagnosis.

That change was enormous.

When disease is divine punishment, the sick person may feel shame.

When disease is understood biologically, the sick person can receive care.

Science did not merely explain disease better.

It made medicine more humane.

It replaced blame with treatment.

Birth Made the Body Sacred

For early human communities, fertility was survival.

A tribe needed children. A farming society needed crops. Herds had to reproduce. Seeds had to grow. Women had to survive childbirth. Infants had to live. Without fertility, the future disappeared.

This made fertility sacred.

The human body seemed mysterious. A woman's body could carry new life. Blood, pregnancy, birth, milk, sexuality, desire, and reproduction were surrounded by awe, fear, rules, rituals, and taboos. The soil also became a kind of body. Seeds entered the earth. Plants emerged. Rain, warmth, and season turned hidden life into food.

It is not difficult to see why human beings connected sex, birth, soil, rain, harvest, and divinity.

Fertility gods, mother goddesses, agricultural rituals, purity rules, sexual taboos, and birth ceremonies grew from the profound dependence of human survival on reproduction and food.

Before biology, conception was mysterious. Before obstetrics, childbirth was dangerous. Before genetics, resemblance between parent and child was wondrous. Before endocrinology, puberty, menstruation, pregnancy, and menopause were deeply puzzling. Before agriculture was scientifically understood, harvest seemed like a miracle repeated by the earth.

Religion entered these mysteries.

It blessed marriage. It regulated sex. It purified menstruation. It protected pregnancy. It prayed for children. It thanked gods for crops. It feared barrenness. It treated birth as sacred because birth was necessary and dangerous.

Again, this was understandable.

But modern knowledge changes the picture. Fertility involves biology, hormones, chromosomes, ovulation, sperm, implantation, placenta, nutrition, disease, and medical care. Crops depend on soil, seed, water, sunlight, pollination, climate, and farming technique. These things may still inspire wonder, but they do not require supernatural explanation.

Science did not make birth less meaningful.

It made it more understandable.

It did not make fertility less precious.

It made it less mysterious.

And when mystery decreases, divine explanation loses ground.

Earthquakes Felt Like Anger Below

When the ground moves beneath the feet, the human sense of safety collapses.

The earth is supposed to be stable. It is the one thing we assume will hold us. But during an earthquake, the ground itself betrays us. Homes crack. Walls fall. Mountains shift. Rivers change. People die under stone and dust.

To early human beings, an earthquake must have felt like cosmic anger.

What power can shake the ground?

What being can break the earth?

What sin can cause such punishment?

Volcanoes were even more dramatic. Fire from a mountain. Smoke darkening the sky. Ash covering fields. Lava moving like glowing death. A mountain that seemed alive, dangerous, and angry.

Before geology, plate tectonics, magma, seismic waves, and the structure of the Earth were known, such events begged for supernatural interpretation. The ground did not seem like a physical crust over a dynamic planet. It seemed like a place where hidden powers lived.

Disaster often creates moral questioning. People do not only ask what happened. They ask why it happened. They ask why some died and others survived. They ask what the victims did. They ask what God wanted. They ask whether the disaster was warning, punishment, sign, or test.

This habit has not disappeared. Even today, some people interpret hurricanes, earthquakes, pandemics, or floods as divine messages. The old pattern remains because the emotional need remains. When disaster is terrible, some people prefer meaning over randomness, even if the meaning is cruel.

But earthquakes are not moral events.

Volcanoes are not sermons.

Tsunamis are not punishments.

They are natural consequences of a restless planet.

This does not reduce the tragedy. It removes the blame. The child crushed under a building was not being punished. The village destroyed by lava did not deserve destruction. The families swept away by flood were not selected by divine anger.

Natural explanation can be painful because it removes cosmic meaning. But it can also be merciful because it removes moral accusation.

A disaster is terrible enough without calling it judgment.

Eclipses Turned Day Into Terror

Most days, the sky follows a rhythm.

The Sun rises. The Sun sets. The Moon changes shape. The stars return. Seasons move in pattern. This rhythm gives comfort.

But sometimes the sky breaks expectation.

The Sun darkens during the day.

A comet appears with a long tail.

A meteor flashes across the night.

A bright object appears where none was expected.

To a prescientific world, unusual sky events could feel terrifying. The sky was the great clock, the great roof, the great mystery above human life. If something strange happened there, it seemed important.

Eclipses especially must have been frightening before they were understood. The Sun, source of light and warmth, appears to disappear. Daylight fades. Birds may grow quiet. Animals may act strangely. People may look upward in terror, wondering whether the world itself is being attacked.

Without astronomy, an eclipse can feel like cosmic danger.

It is not surprising that such events were interpreted as omens. A comet could be linked to the fall of a king. An eclipse could be seen as divine warning. A meteor could become a sign from heaven.

But astronomy changed all this.

An eclipse is not a message. It is an alignment.

A comet is not a warning. It is an icy body moving through space.

A meteor is not divine writing. It is matter entering the atmosphere.

The power of astronomy was not only that it explained these events. It predicted them. Prediction is one of science's greatest strengths. A priest may interpret an eclipse after it happens. Astronomy can predict it before it happens.

That difference matters.

To explain afterward is easy.

To predict accurately is powerful.

When science predicts what religion once feared, the old supernatural interpretation becomes weaker. The sky is no longer a screen for divine messages. It becomes a lawful physical reality.

The mystery remains beautiful, but it no longer needs fear.

Omens Made Coincidence Feel Like Communication

An omen is an attempt to turn uncertainty into communication.

A bird flies in a certain direction. A dream comes before a battle. A comet appears before a ruler dies. A child is born during a storm. A crop fails after a ritual is neglected. A person becomes sick after offending an elder.

The mind connects events.

This happened, then that happened. Therefore, perhaps the first caused the second. Perhaps the universe is speaking. Perhaps the gods are warning us. Perhaps nothing is accidental.

Omens are powerful because they give the feeling of hidden order. They suggest that the world is not random. Every event may mean something. Every unusual occurrence may carry a message.

But the human mind is very good at finding connections that may not exist.

We remember the dream that seemed to come true. We forget the many dreams that did not. We remember the prayer followed by recovery. We forget the prayers followed by death. We remember the comet before the king's death. We forget the comets that passed without consequence.

This is not stupidity. It is ordinary human psychology. The mind prefers meaningful patterns. It does not easily accept coincidence.

Religious systems often grow around this habit. Omens become traditions. Traditions become rules. Rules become authority. Authority becomes power.

Someone claims to read the signs.

Someone claims to know what the gods want.

Someone becomes interpreter of the invisible.

This gives religious leaders influence. If people believe nature is full of messages, then the person who claims to read those messages becomes important. He can advise kings, frighten villagers, bless wars, condemn enemies, explain disasters, and demand obedience.

The omen is not just a belief. It can become a system of control.

Science weakens omens by explaining coincidence, probability, natural causes, and human bias. It does not remove all uncertainty. But it gives us a better way to ask questions.

Not, "What is the universe trying to tell me?"

But, "What evidence shows that these events are truly connected?"

That question changes everything.

From Pleasing Nature to Understanding Nature

Before science, human beings tried to manage nature by pleasing invisible powers.

They prayed for rain.

They sacrificed for harvest.

They confessed for illness.

They purified after disaster.

They interpreted eclipses.

They feared comets.

They begged gods to stop plague.

They buried objects with the dead.

They performed rituals to protect childbirth.

They did these things because they wanted influence over forces they did not understand.

Science changed the goal.

Instead of asking how to please the storm, science asked what causes the storm.

Instead of asking which god sent disease, medicine asked what causes infection.

Instead of asking why thc earth was angry, geology asked why the ground moved.

Instead of asking what an eclipse meant, astronomy asked how the Sun, Moon, and Earth were aligned.

Instead of asking what ritual would bring harvest, agriculture asked what soil, seed, water, climate, and labor could do.

This was a revolution in human thought.

Appeasement gave way to understanding.

Fear gave way to investigation.

Ritual gave way, slowly and imperfectly, to method.

That does not mean science removed wonder. A scientific view of nature can be full of wonder. The birth of a child, the structure of a cell, the power of a storm, the motion of planets, the deep time of rocks, and the light from distant stars are all astonishing.

But wonder does not require worship.

A natural explanation does not make the world empty. It makes the world clearer.

Religion once placed gods inside thunder, disease, rain, fertility, earthquakes, and eclipses. Science did not destroy those things. It explained them. And as explanation grew, the gods that lived inside those mysteries had less work to do.

Nature remained powerful.

But it no longer needed to be divine.

Chapter Summary

Before science, nature was frightening, unpredictable, and often deadly. Thunder, lightning, drought, flood, disease, childbirth, earthquake, volcano, eclipse, comet, and plague all seemed to demand explanation. Without knowledge of natural causes, human beings often interpreted these events as divine anger, blessing, curse, punishment, omen, or supernatural message.

This chapter has shown that the religious interpretation of nature was understandable. It gave people a sense of meaning, control, and emotional relief in a dangerous world. But understandable does not mean true. Modern science has shown that storms, disease, earthquakes, eclipses, and droughts follow natural causes, not moral intentions.

Nature is powerful, but power is not personality.

Nature can destroy us, but destruction is not judgment.

The world can frighten us, but fear is not evidence of God.

Key Points

Before science, natural forces often seemed personal, intentional, and morally meaningful.

Thunder, lightning, drought, disease, fertility, earthquakes, volcanoes, eclipses, and comets helped create supernatural explanations.

Rituals and prayer gave people a feeling of control over frightening natural events.

Disease was often interpreted as curse, punishment, impurity, or possession before germ theory and modern medicine.

Natural disasters are physical events, not moral judgments.

Omens arise when the human mind connects events and interprets coincidence as message.

Science replaced appeasement with understanding.

Nature remains awe-inspiring, but it does not require divine explanation.

Transition to the Next Chapter

Death frightened the human heart. Nature frightened the human body. But there was another powerful force behind the creation of God: the human mind itself.

The mind searches for patterns. It imagines causes. It detects agency. It sees intention in movement and meaning in coincidence. These habits helped our ancestors survive, but they also made gods easier to imagine.

The next chapter examines the pattern-seeking brain, and why the human mind is so ready to believe in invisible powers.

Sources

Boyer, Pascal. Religion Explained: The Evolutionary Origins of Religious Thought. Basic Books, 2001.

Burkert, Walter. Creation of the Sacred: Tracks of Biology in Early Religions. Harvard University Press, 1996.

Hume, David. The Natural History of Religion. 1757.

Sagan, Carl. The Demon-Haunted World: Science as a Candle in the Dark. Random House, 1995.

Tylor, Edward B. Primitive Culture. 1871.

Chapter 4: The Pattern-Seeking Brain

The Mind That Could Not Leave Mystery Alone

Human beings do not simply see the world.

We interpret it.

A sound in the dark is not just a vibration in the air. We wonder what caused it. A shadow near the trees is not just light and angle. We ask whether something is there. A sudden illness is not merely an event. We ask why it happened, why now, and why this person. A coincidence rarely remains a coincidence for long. The mind begins to connect, explain, and give meaning.

This habit helped our ancestors survive.

A rustle in the grass might be wind. It might also be a snake. A moving shape in the distance might be a branch. It might also be a predator. A broken twig might mean nothing. It might also mean an enemy is nearby.

The mind that quickly asked, "Who is there?" had a better chance of survival than the mind that waited too long.

So the brain became a pattern-seeking organ.

It looked for footprints, faces, danger, intention, memory, threat, opportunity, and cause. It looked for hidden agents behind visible signs. That ability helped human beings hunt, avoid predators, recognize allies, remember enemies, and survive in dangerous environments.

But the same ability also made gods easier to imagine.

If the mind is always searching for a hidden cause, thunder may become a voice. Disease may become punishment. Coincidence may become a sign. A dream may become a message. A shadow may become a spirit. A narrow escape may become divine protection. A tragedy may become God's will.

The human brain is brilliant, but it is not perfect. It was shaped first for survival, not perfect truth. It often prefers a quick explanation to no explanation. It often prefers a meaningful story to meaningless chance. It often prefers an invisible agent to a blind natural process.

This chapter is about that mind: the mind that created tools, language, fire, art, medicine, mathematics, and science, but also spirits, demons, omens, miracles, curses, and gods.

Patterns Saved Lives, but Also Created Illusions

Pattern recognition is one of the great powers of the human brain.

A hunter sees tracks in the mud and knows an animal passed. A mother hears a change in her child's breathing and senses danger. A farmer watches clouds, wind, soil, and insects and guesses what weather may come. A physician listens to symptoms and recognizes a pattern of disease. A scientist studies repeated observations and discovers a law of nature.

Without pattern recognition, there would be no survival, no medicine, no engineering, no farming, no science, and no civilization.

The problem is not that the brain sees patterns.

The problem is that it sometimes sees patterns too eagerly.

The brain may connect two events that are not truly connected. A person prays, and later rain comes. A man dreams of an accident, and days later someone is injured. A woman wears a charm, and her sick child recovers. A village performs a ritual, and the harvest improves. A soldier carries a sacred object and survives battle.

The mind remembers the connection.

Prayer came before rain.

Dream came before accident.

Charm came before recovery.

Ritual came before harvest.

Sacred object came before survival.

A story is born.

But many prayers do not bring rain. Many dreams do not come true. Many charms do not heal. Many rituals do not change harvests. Many soldiers carrying sacred objects die. The brain often remembers the striking matches and forgets the ordinary misses.

This is not rare.

It is human.

We are drawn to patterns because patterns help us feel that the world is understandable. A pattern gives comfort. A pattern gives control. A pattern tells us what to do next. Even a false pattern can feel better than uncertainty.

This helps explain why religious belief can grow from ordinary human thinking. The believer may not be lying. The believer may be remembering selectively, interpreting emotionally, and connecting events through meaning rather than evidence.

A pattern can save your life.

But a false pattern can create a god.

The Rustle in the Grass

Imagine two early humans walking near tall grass.

Both hear a sudden rustle.

One thinks, "It is probably only the wind," and continues walking.

The other thinks, "It may be a predator," and steps back.

If it is only the wind, the cautious person loses little. Perhaps he feels foolish. Perhaps he wastes a moment. But if it is a predator, the cautious person survives while the careless person may die.

This is the logic of false alarms.

In a dangerous world, it is often safer to assume agency than to ignore it. Better to mistake wind for a predator than a predator for wind. Better to imagine danger where none exists than to miss danger when it is real.

Over many generations, this kind of thinking may have shaped the human mind. We became sensitive to possible agents: animals, enemies, watchers, helpers, attackers, ancestors, spirits, and gods. We became quick to ask, "Who is behind this?"

This habit is useful in the forest.

It is less useful when explaining the universe.

A storm has no mind. A virus has no intention. An earthquake has no anger. A drought has no moral opinion. A genetic disease does not know the child it harms. A comet does not announce political change. A dream does not prove that the dead are speaking.

But the same brain that once protected us from predators may still search for agents behind impersonal events.

This is one reason gods feel natural to the human mind. A god is the ultimate hidden agent. God explains what we cannot see. God acts behind the visible world. God knows, chooses, watches, punishes, protects, speaks, and plans. God gives agency to nature and intention to history.

The religious mind often says, “Someone must be behind this.”

Science asks a different question: “What is the evidence that someone is behind this?”

That difference is enormous.

The Invisible Watcher

Human beings are especially good at thinking about other minds.

We guess what others know, want, fear, intend, hide, remember, and plan. This ability is essential for social life. Without it, we could not cooperate, deceive, trust, teach, love, lead, or protect ourselves.

A child quickly learns that other people have thoughts. A friend may be angry but silent. A stranger may be kind or dangerous. A leader may speak one thing and intend another. A mother may know what the child is hiding. A rival may plan revenge.

Human life depends on reading invisible mental states.

We cannot see another person’s intention directly, but we infer it. We cannot see desire, jealousy, guilt, loyalty, or deception, but we feel their presence through behavior. This ability is so important that the human mind uses it constantly.

Religion may have taken this ordinary social skill and expanded it beyond visible human beings.

If people have invisible minds, perhaps nature has invisible minds.

If a father can watch, judge, and punish, perhaps an ancestor can do the same.

If a king can command from far away, perhaps a god can command from heaven.

If a mother can know when a child is afraid, perhaps a divine being can know the heart.

If a human enemy can act in secret, perhaps demons and spirits can do the same.

This is how the social mind may have helped create religious belief. Human beings already lived in a world of invisible intentions. Religion filled the larger world with invisible persons.

God became a mind behind the world.

This helps explain why religious believers often speak of God in deeply personal language. God loves. God listens. God wants. God commands. God forgives. God becomes angry. God tests. God watches. God has a plan.

These are not scientific properties. They are social properties. They are the language of human relationships, enlarged and projected onto the universe.

The same mind that understands other people may have helped invent gods.

Coincidence Dressed as Destiny

Coincidence is one of the strongest fuels of religious belief.

A person prays for help, and the phone rings.

A mother worries about her son, and later learns he was in danger.

A man thinks of an old friend, and the friend calls that day.

A patient improves after a blessing.

A traveler misses a bus, and later hears the bus had an accident.

A family asks for a sign, and a bird appears at the window.

To the human mind, such moments can feel charged with meaning. They do not feel random. They feel arranged, directed, personal. They feel like the universe is speaking.

But the world is full of events, and the brain is always connecting them.

Most thoughts do not become reality. Most prayers are followed by ordinary outcomes. Most dreams do not predict anything. Most birds at windows are only birds at windows. Most coincidences are noticed because they stand out from thousands of unnoticed non-coincidences.

The mind remembers the hit and forgets the misses.

If someone dreams of a death and a death occurs, the dream becomes a story. But what about the many dreams that did not come true? What about the many anxieties that did not predict anything? What about the many prayers that were not followed by recovery?

Religious belief often grows by preserving meaningful coincidences and quietly ignoring the rest.

This does not happen only in religion. It happens in gambling, superstition, sports, politics, medicine, and daily life. A gambler wears a lucky shirt. An athlete repeats a pregame ritual. A patient credits a supplement after improvement. A person avoids a road because "something felt wrong."

The mind is hungry for connection.

Science responds by asking for more than a memorable story. It asks: how often does this happen? Compared with what? Is there a control group? Can it be repeated? Can it be measured? Can the explanation predict future events better than chance?

That is why science is so important. It protects us from the mind's eagerness to turn coincidence into truth.

The Comfort of a Hidden Plan

Random suffering is one of the hardest things for human beings to accept.

A child dies. A young mother gets cancer. A careful driver is killed by a drunk driver. A kind man loses everything in a flood. A family is destroyed by an earthquake. A newborn has a severe genetic disorder.

The human heart asks, "Why?"

Often, the honest answer is painful: there may be no personal reason. The cause may be biological, environmental, genetic, accidental, or natural. The event may be terrible without being intended.

Religion often offers another answer.

It says there is a plan.

It says God has a reason.

It says the suffering is a test.

It says the pain will be rewarded.

It says the event is part of a mystery beyond human understanding.

For many people, this is comforting. Purpose is easier to bear than randomness. A planned sorrow may feel less frightening than meaningless sorrow. If someone is in control, then perhaps life is not as fragile as it appears.

But this comfort comes at a cost.

If suffering is part of a divine plan, then terrible questions arise. What kind of plan needs childhood cancer? What kind of purpose requires earthquakes that crush families? What kind of love allows animals to suffer for millions of years before humans existed? What kind of wisdom needs hunger, disease, and terror?

Those questions belong more fully to Book 6, The Failure of the Personal God. Here, the point is simpler: the human need for purpose is one reason God becomes believable.

The mind does not like blind suffering. It tries to place suffering inside a story. Religion supplies the largest story of all.

But a story that comforts us is not automatically true.

Sometimes pain has a cause, but no purpose.

That may be hard to accept. But it may also be more honest, and in some ways more humane. If disease is not a test, we should treat it. If disaster is not punishment, we should rescue people. If suffering is not God's mysterious plan, then reducing suffering becomes our responsibility.

The search for hidden purpose created many gods.

Science asks us to look for causes instead.

Faces in the Clouds

The human brain is especially quick to see faces.

A few shapes on a wall may look like eyes and a mouth. A pattern in wood may resemble a human profile. A cloud may look like an animal. A shadow may look like a figure. A rock formation may look like a face.

This is not weakness. It is part of being human.

Faces matter. The ability to recognize faces quickly helps infants bond, helps adults read emotion, and helps groups live together. A face tells us if someone is angry, afraid, kind, deceptive, sick, or dangerous. We are built to notice faces because faces are socially important.

But this ability can also mislead us.

We can see faces where there are no faces. We can see intention where there is no intention. We can see meaning where there is only shape, light, chance, and expectation.

Religious imagery often grows from this mental habit. People may see divine figures in clouds, stains, tree bark, shadows, or patterns of light. They may interpret a natural shape as a sign. The emotional reaction can be strong because the brain treats the image as meaningful.

The same process can happen without visual images. A person may hear a message in a random song, see divine timing in an ordinary delay, or interpret a sudden feeling as God speaking.

The brain is not only a receiver of reality. It is also a maker of meaning.

This does not mean all meaning is false. Human beings can create real meaning through love, work, memory, art, service, and moral responsibility. But when meaning is mistaken for external evidence, belief can move beyond what the facts support.

Seeing a face in a cloud does not prove someone is in the cloud.

Feeling meaning in an event does not prove the event was arranged by God.

Belief Arrives Before Testing

The scientific method is not natural to the human mind.

Science asks us to slow down. It asks us to test. It asks us to doubt first impressions. It asks us to compare evidence. It asks us to separate what we want to be true from what is supported. It asks us to accept correction. It asks us to change our minds.

That is difficult.

The human mind usually works faster. It believes, explains, justifies, and defends. It feels first and reasons later. It often protects what is familiar. It trusts the group. It remembers supporting evidence and forgets opposing evidence. It prefers certainty to doubt.

Religion often benefits from this.

A child hears about God before the child understands evidence. A family repeats prayers before the child can compare religions. A community teaches sacred stories before the child can read history. Fear of punishment may enter the mind before critical thinking develops. By the time the adult begins to question, belief may already feel obvious.

The mind then does what minds often do: it defends what it already holds.

It notices answered prayers.

It explains unanswered prayers.

It remembers moments of comfort.

It ignores contradictions.

It calls doubt a test.

It calls coincidence a sign.

It calls inherited belief truth.

This is not unique to religion. Human beings do this in politics, nationalism, family loyalty, medical myths, conspiracy theories, and personal identity. We defend beliefs that feel important to us.

Science developed because ordinary thinking is not enough. We needed methods to protect us from ourselves. Measurement, repeated testing, peer criticism, controlled observation, and willingness to revise are not perfect, but they are better than intuition alone.

Religion often asks the believer to trust.

Science asks the thinker to check.

That difference is one reason science gradually made God less necessary as an explanation.

Why the Invisible Feels Plausible

One might ask: why did humans believe in invisible beings so easily?

The answer is that much of human life already depends on invisible things.

We cannot see thoughts, but we know people think.

We cannot see love, but we see its effects.

We cannot see anger inside the mind, but we see the face, voice, and action.

We cannot see intention directly, but we infer it from behavior.

We cannot see memory, but we know it shapes action.

So the idea of invisible agency did not feel strange. Human beings already lived among invisible mental realities. Religion extended that idea into nature and the cosmos.

If my mother can love me without love being visible, perhaps God can love me invisibly.

If my enemy can secretly plan against me, perhaps demons can do the same.

If my father can judge my behavior, perhaps ancestors or gods can judge me.

If a leader can command from far away, perhaps a divine king can command from heaven.

Invisible gods were not psychologically difficult to imagine because invisible minds were already part of human social life.

But there is an important difference.

Human minds are connected to human bodies. We infer a person's thoughts because the person acts, speaks, moves, responds, remembers, and interacts. We can test human agency in ordinary ways.

With gods, the evidence is far less clear. The believer may see God's action in nature, coincidence, feeling, or tradition, but those things can often be explained without God. The invisible agent becomes difficult to test. Whatever happens can be interpreted as divine will. Recovery is God's mercy. Death is God's plan. Silence is God's mystery. Disaster is God's warning. Absence is God's hiddenness.

A belief that can explain every outcome may actually explain very little.

Science asks for explanations that risk being wrong. A good explanation must meet reality. It must be open to correction. It must tell us what we should expect if it is true and what would count against it.

Invisible gods often escape that kind of testing.

That makes them emotionally powerful, but scientifically weak.

The Brain That Made Gods Also Made Science

The pattern-seeking brain is not an enemy.

It is part of our greatness.

Because of this brain, human beings learned to track animals, predict seasons, recognize disease patterns, create language, tell stories, build tools, form societies, and eventually develop science. The same mind that imagined gods also discovered mathematics, music, surgery, astronomy, and the structure of DNA.

The problem is not the brain's search for meaning. The problem begins when the search for meaning becomes disconnected from evidence.

Early humans needed quick guesses. Science needs careful testing.

Early humans needed stories to survive fear. Science needs explanations that can survive criticism.

Early humans saw agency in the unknown. Science asks whether agency is actually there.

This is the great movement from religion to science.

Religion often begins with the feeling that something must be behind the event.

Science begins with the question: what is actually causing the event?

That shift changed human history.

The storm became weather.

The plague became infection.

The eclipse became alignment.

The earthquake became geology.

The dream became brain activity.

The voice in the dark became wind, memory, fear, or imagination.

Slowly, the invisible beings that once filled the world had less work to do.

The pattern-seeking brain created gods because it was trying to survive in mystery. The scientific mind began when human beings learned to question their own patterns.

That may be one of the greatest steps in human history.

Chapter Summary

The human brain is built to search for patterns, causes, and hidden agents. This ability helped our ancestors survive in dangerous environments. It allowed them to detect predators, understand social behavior, remember danger, and respond quickly to uncertainty.

But the same ability also made supernatural belief easier. The mind may see intention behind storms, punishment behind disease, messages in coincidence, purpose in suffering, and invisible agents behind natural events. Religious belief often grows from ordinary human mental habits: pattern recognition, agency detection, fear, memory, social thinking, and the search for meaning.

This does not mean religious believers are dishonest. It means they are human. The brain can create sincere beliefs that feel deeply real but still go beyond the evidence. Science became necessary because intuition alone is not enough. We needed methods that could test our stories, correct our errors, and separate comforting patterns from reliable truth.

The brain that imagined God also gave us the power to question God.

That may be the beginning of freedom.

Key Points

The human brain naturally searches for patterns, causes, and hidden meaning.

Pattern recognition helped early humans survive, but it can also create false beliefs.

Agency detection makes people quick to imagine a hidden actor behind events.

Coincidences can feel like signs when the mind connects events emotionally.

The brain often remembers meaningful hits and forgets ordinary misses.

Religious belief can grow from normal human mental habits, not from dishonesty.

Science helps protect us from false patterns by demanding evidence, testing, and correction.

The mind that created gods also created science, but science begins when the mind learns to question itself.

Transition to the Next Chapter

The pattern-seeking brain made human beings ready to see hidden agents behind the world. But another experience made the invisible world feel even more real: dreams.

In dreams, the dead return. The body lies still, but the dream-self travels. Voices speak. Faces appear. Places change. The impossible feels real until morning.

For early humans, dreams may have opened the door to spirits, ancestors, souls, and unseen worlds.

The next chapter examines how dreams, spirits, ancestors, and the invisible world helped shape belief in God.

Sources

Barrett, Justin L. Why Would Anyone Believe in God? AltaMira Press, 2004.

Boyer, Pascal. Religion Explained: The Evolutionary Origins of Religious Thought. Basic Books, 2001.

Guthrie, Stewart Elliott. Faces in the Clouds: A New Theory of Religion. Oxford University Press, 1993.

Kahneman, Daniel. Thinking, Fast and Slow. Farrar, Straus and Giroux, 2011.

Shermer, Michael. The Believing Brain: From Ghosts and Gods to Politics and Conspiracies. Times Books, 2011.

Chapter 5: Dreams, Spirits, Ancestors, and the Invisible World

When the Dead Returned in the Dark

A man dies.

His body is buried. His voice is gone. His place near the fire is empty. The tools he used remain where he left them. His children still expect to see him enter the shelter. His wife still turns toward footsteps that are not his. His friends remember his laugh, his anger, his advice, his strength, and his final silence.

Days pass.

Then one night, he returns.

Not in daylight. Not in the village. Not in the ordinary world of touch, weight, and waking sight. He returns in a dream.

He speaks. He walks. He smiles. He gives advice. He looks alive. For a few moments, death seems to have failed.

Then morning comes.

The dreamer wakes. The body is still gone. The grave is still there. But something has changed. The dead man felt present. His voice was heard. His face was seen. The dream did not feel like an idea. It felt like an encounter.

Before modern science, what could this mean?

Today we may speak about REM sleep, memory, grief, neural activity, emotional processing, and the brain's ability to create vivid inner worlds. But early human beings had none of this knowledge. They did not have sleep laboratories, brain scans, neurology, psychology, or modern theories of dreaming. To them, a dream was not easily dismissed as "only a dream." It was an experience, and powerful experiences demand interpretation.

Dreams may have been one of the first doors through which human beings entered the idea of an invisible world.

In dreams, the dead returned.

In dreams, the body stayed still, but the self traveled.

In dreams, absent people appeared.

In dreams, impossible things felt real.

In dreams, another world seemed to open.

For a mind without neuroscience, dreams could easily suggest that visible reality is not the whole of reality. The waking world was one realm. The dream world seemed to be another. And if the dead could appear there, perhaps they were not fully dead.

This idea did not require formal theology. It required only grief, sleep, memory, imagination, and the refusal of love to accept disappearance.

The Sleeping Body and the Wandering Self

Dreams create a strange puzzle.

The body lies still, but the dreamer moves.

The sleeping person remains near the fire, yet in the dream he walks through a forest, climbs a mountain, speaks to a dead relative, runs from danger, or visits a place far away.

To modern science, this is not mysterious in the same way. The brain can create scenes, emotions, sounds, faces, and movement while the body rests. But to early human beings, the experience could suggest something else.

Perhaps there is a second self.

Perhaps the body sleeps, but the inner person travels.

Perhaps the visible body is not the whole person.

This idea appears in many forms across human cultures. Some traditions imagine a soul. Some imagine a spirit. Some imagine a breath-self, shadow-self, dream-self, or double. The names differ, but the central thought is similar: something inside the human being may separate from the body.

Dreams make this feel natural.

A hunter dreams of a distant valley.

A mother dreams of her dead child.

A healer dreams of entering the spirit world.

A sick person dreams of being taken by invisible beings.

A young person dreams of an animal that becomes sacred.

The dream does not need to be logical. It needs to be vivid. A single powerful dream can remain in memory for years. The dreamer wakes with emotion still alive in the body: fear, comfort, longing, guilt, warning, or peace. The feeling itself becomes part of the evidence.

Human beings often trust intense experience. If something shakes us deeply, we assume it must mean something. The stronger the feeling, the stronger the belief may become.

But emotional intensity is not proof of external reality.

A dream may reveal grief. It may reveal fear. It may reveal desire. It may reveal memory. It may reveal the brain's ability to create living scenes from inner material. But it does not automatically reveal a supernatural world.

Still, before science, the dream-self was one of the most natural paths toward belief in the soul.

A World Alive With Spirits

If dreams suggested that the self could leave the body, and death suggested that something invisible had departed, then the world could easily become filled with spirits.

The dead might remain near.

Animals might have spirits.

Trees might have spirits.

Rivers might have spirits.

Mountains might have spirits.

Storms might have spirits.

Objects might carry power.

Places might hold memory.

This kind of thinking is often called animism: the belief that the world is alive with spirits, souls, or unseen agencies. It is one of the oldest and most widespread forms of religious imagination.

Animism makes human sense.

Early people lived close to nature. They depended on animals, plants, rivers, seasons, fire, and land. A forest could feed or kill. A river could nourish or drown. An animal could become food or predator. A tree could give shelter, fruit, wood, shade, and fire. The world was not an abstract physical system. It was a living field of relationship.

To imagine spirit in nature was to treat nature as a society.

The mountain was not only rock.

The river was not only water.

The animal was not only meat.

The forest was not only trees.

Each could be approached, respected, feared, thanked, or offended. Hunters might honor the animal they killed. Farmers might ask the land for fertility. Families might speak to ancestors. Villagers might avoid places believed to be spiritually dangerous.

This may sound strange to the modern scientific mind, but it reveals a deep human habit. We understand persons before we understand systems. We understand anger before we understand atmospheric pressure. We understand hunger before we understand ecology. We understand intention before we understand biology.

So early religion often turned the world into a community of invisible persons.

Science would later describe rivers through hydrology, forests through ecology, disease through pathology, storms through meteorology, and mountains through geology. But before these explanations existed, spirits gave the world a human-like structure.

The world became social.

And a social world requires respect, ritual, and rules.

Ancestors Became the First Immortals

Among the most powerful spirits were ancestors.

The dead did not disappear emotionally. They remained in memory, story, duty, land, custom, and family identity. A dead father could still command obedience through remembered words. A dead mother could still shape a child's conscience. A dead elder could still guide decisions through tradition. A dead ruler could still legitimize power.

In this way, the dead remained socially alive.

Ancestor belief may have grown naturally from this emotional and social reality. The living felt the dead were still present because, in many ways, they were. They lived in memory. They lived in inherited rules. They lived in names, objects, graves, songs, land, and family stories.

A family might ask: what would our ancestors want?

A tribe might say: this land belongs to us because our ancestors are buried here.

A child might be told: your grandfather is watching.

A leader might claim: the ancestors have chosen this path.

Once the dead were imagined as continuing beings, they could become protectors, judges, witnesses, or sources of blessing. They could punish disrespect. They could protect descendants. They could appear in dreams. They could demand ritual attention.

This gave religion emotional depth and social power.

Honoring ancestors connected people to the past. It strengthened family loyalty. It gave authority to tradition. It made the living feel watched by those who came before. It also made death less final. The dead remained part of the community, not merely as memory, but as invisible presence.

We must be careful here. Ancient burial practices and rituals do not allow us to know exactly what early people believed. We cannot enter their minds. But across human cultures, the dead often became central to religious imagination.

That is understandable.

The dead leave the body, but not the heart.

Religion gave that continuing emotional presence a supernatural form.

Shadows, Reflections, and the Double Self

Dreams were not the only experiences that may have encouraged belief in an invisible self.

A human being sees a shadow follow the body. The shadow moves when the body moves. It changes shape. It appears and disappears with light. A reflection appears in water: a face, a body, a second image. The person looks into the surface and sees another self looking back.

To modern people, shadows and reflections are optical phenomena. But long before formal science, they may have carried deeper mystery.

What is the shadow?

What is the reflection?

Is it part of the person?

Can it be harmed?

Can it leave?

Can it be captured?

Many cultures have treated shadows, names, images, breath, blood, hair, and footprints as connected to a person's inner being. This may seem superstitious, but it follows a natural human idea: the visible trace is linked to the invisible self.

Even today, we feel something similar. A photograph of a loved one is not merely paper. A wedding ring is not merely metal. A child's toy after death is not merely cloth or plastic. A handwritten note from a dead parent is not merely ink. Objects can carry emotional presence. They become more than their physical parts because memory attaches to them.

The ancient mind may have taken this further.

A shadow, reflection, name, or personal object could feel like part of the person. If the person died, some trace might remain. If a dream image appeared, perhaps the person had another form. If breath stopped, perhaps life had flown away. If a reflection appeared, perhaps the self had a visible double.

Such experiences may have helped build the idea that human beings are not only bodies.

Science gives natural explanations. Shadows are produced when light is blocked. Reflections come from light bouncing from surfaces. Breath is gas exchange. Blood is tissue and fluid. Hair and names carry biological and social meaning, not supernatural essence.

But before science, the boundary between body, image, name, breath, and spirit may have felt less clear.

The self seemed to have many signs.

Religion gathered those signs and turned them into soul.

Visions Gave Authority to the Unseen

Some people do not merely dream at night. They enter unusual states while awake.

Across many cultures, shamans, healers, prophets, mystics, or visionaries have claimed to enter spirit worlds, speak with ancestors, receive messages, heal illness, locate lost things, predict events, or battle unseen forces.

These experiences may involve chanting, drumming, fasting, dancing, isolation, sleep loss, grief, illness, meditation, sensory overload, sensory deprivation, or psychoactive plants. Such practices can change perception, emotion, time sense, body awareness, and the feeling of self.

To the person having the experience, it can feel deeply real.

A voice may be heard.

A presence may be felt.

A vision may appear.

The body may shake.

Time may seem altered.

The self may feel outside the body.

A dead person, animal guide, spirit, angel, demon, or god may seem present.

Before neuroscience, such experiences could easily be interpreted as contact with an invisible world. The visionary became important because he or she seemed to cross the border between ordinary life and the spirit realm.

This gave the shaman, prophet, or mystic social power.

If most people fear spirits, the person who claims to deal with spirits becomes valuable. He can heal, interpret, warn, bless, curse, guide, or command. He becomes a bridge between the visible and invisible worlds.

Some such people may have been sincere. Some may have had unusual psychological experiences. Some may have been skilled observers of human emotion. Some may have used performance, suggestion, and authority. Some may have deceived. Human history usually contains all possibilities.

The key point is this: altered states can feel like revelation, but feeling is not proof.

The brain is capable of producing extraordinary experiences. A person can hear voices, see visions, feel presence, lose body boundaries, experience ecstasy, or feel terrifying certainty. These states may be meaningful to the person, but they do not automatically prove that the experience came from outside the brain.

A vision may be sincere and still be internal.

A trance may be powerful and still be natural.

A revelation may transform a life and still fail as evidence of God.

The Invisible Became More Real Than the Visible

One of religion's most remarkable achievements is that it made invisible things feel more important than visible things.

The visible body dies, but the invisible soul lives.

The visible world changes, but the invisible realm is eternal.

The visible king rules briefly, but the invisible God rules forever.

The visible action matters, but the invisible intention matters even more.

The visible life is short, but the invisible afterlife is endless.

This shift gave religion enormous power.

If the invisible world is more real than the visible world, then the priest, prophet, scripture, ritual, and doctrine become more important than ordinary evidence. The believer may be told not to trust the senses, not to trust doubt, not to trust reason, not to trust the body, not to trust the world. The invisible becomes the higher truth.

This can be comforting, but it can also be dangerous.

When invisible claims become stronger than visible evidence, almost anything can be defended. A failed prophecy can be reinterpreted. An unanswered prayer can be called mystery. A disaster can be called divine plan. A contradiction can be called deeper truth. A cruel doctrine can be called holy. A leader's command can be called God's will.

This is why the invisible world must be examined carefully.

Science does not deny that invisible things exist. Many real things are invisible to ordinary sight: germs, atoms, radiation, gravity, air pressure, magnetic fields, electrical signals, and DNA. But science does not accept them merely because someone feels them or claims them. It tests their effects. It measures them. It predicts with them. It builds evidence around them.

The invisible is not the problem.

Unsupported invisible claims are the problem.

This distinction is crucial. Science has discovered many invisible realities. Religion often asks us to accept invisible beings without comparable evidence.

A god who cannot be seen, measured, tested, predicted, or distinguished from imagination is not like gravity or germs. Gravity and germs leave reliable evidence. They can be studied. They can be tested. They can be used to predict and explain. God-claims rarely behave that way.

The invisible world of religion may be emotionally powerful, but emotional power is not enough.

Inner Experience Became Outer Truth

A major mistake in religious thinking is the movement from inner experience to outer reality.

A person feels peace and concludes God is present.

A person hears a voice and concludes God spoke.

A person dreams of the dead and concludes the dead survive.

A person feels watched and concludes a divine observer exists.

A person has a vision and concludes heaven opened.

A person senses meaning and concludes the universe has a plan.

But inner experience does not automatically prove outer reality.

The mind can create peace.

The brain can produce voices.

Memory can create dreams.

Fear can create presence.

Expectation can shape perception.

Culture can give language to experience.

This does not mean the experiences are unimportant. They may be deeply important. They may help a person endure grief, change behavior, find courage, or feel connected. But the question remains: what caused the experience?

The believer may say God.

The skeptic may say brain, memory, emotion, culture, and expectation.

Which explanation is stronger?

A good explanation should not merely repeat the experience. It should explain why different people in different cultures have different visions. It should explain why religious experiences often match what people already believe. It should explain why a Hindu may experience Hindu imagery, a Christian may experience Jesus or Mary, a Muslim may experience Islamic themes, and a person in another culture may experience ancestors, spirits, or local deities.

This cultural shaping is important.

If religious experiences came from one clear external God, we might expect more consistency. Instead, religious experience often wears the clothing of culture.

The mind has the experience.

Culture gives it a name.

Religion gives it authority.

Another World Was Built From Human Experience

The invisible world probably did not begin as one complete idea. It likely grew slowly from many human experiences: dreams, death, breath, shadows, reflections, hallucinations, trance, grief, fear, and memory.

One experience alone may not create religion. But together, they build a powerful picture.

The dead appear in dreams.

The body dies, but something seems to leave.

The shadow follows the body.

The reflection looks like a second self.

The shaman enters trance.

The grieving feel a presence.

The sick hear voices.

The frightened see signs.

The tribe remembers ancestors.

The story grows.

Over time, the visible world becomes surrounded by an invisible world. The living are not alone. The dead watch. Spirits move. Ancestors judge. Gods speak. Dreams carry messages. Rituals open doors. Sacred places hold power.

This invisible world gave human beings comfort, fear, order, identity, and explanation. It helped them live with death, uncertainty, disease, and danger.

But it also created a problem.

Once people believed in an invisible world, they needed interpreters. Who can speak with spirits? Who can read dreams? Who can explain visions? Who can know what the ancestors want? Who can tell whether a dream is warning, memory, or nonsense? Who can protect the tribe from unseen forces?

This need gave rise to spiritual authority.

The next steps toward organized religion had begun.

Chapter Summary

Dreams, death, shadows, reflections, altered states, grief, and memory may have helped early humans imagine an invisible world. In dreams, the dead returned. The sleeping body remained still, but the dream-self seemed to travel. The dead continued in memory and appeared in inner experience. Shadows and reflections suggested a double self. Trance and vision gave some people the feeling that they had entered another realm.

These experiences made belief in spirits, ancestors, souls, and invisible powers emotionally natural. They did not prove a supernatural world, but they helped make such a world believable. Religion often grew by interpreting inner experiences as outer realities.

The key distinction is this: an experience can feel real without proving that its interpretation is true. A dream of the dead may reveal grief and memory. A vision may reveal the brain's power. A trance may reveal altered consciousness. But none of these automatically proves the existence of spirits or God.

The invisible world may have begun not in heaven, but in the human mind.

Key Points

Dreams may have strongly encouraged belief in spirits, souls, and life after death.

The sleeping body and traveling dream-self may have suggested a separable soul.

Ancestor belief helped keep the dead emotionally and socially present among the living.

Shadows, reflections, breath, names, and personal objects may have contributed to belief in an invisible self.

Trance, visions, and altered states gave spiritual specialists authority.

The invisible world became emotionally powerful because it answered grief, fear, memory, and longing.

Science accepts invisible realities only when they leave reliable evidence.

Inner experience does not automatically prove external supernatural reality.

Transition to the Next Chapter

Once human beings imagined invisible agents, those agents began to take human form.

Gods became fathers, kings, judges, warriors, protectors, lawgivers, and rulers. They did not appear as neutral forces. They carried human emotions, human authority, human anger, human love, and human demands.

This is one of the strongest clues that gods were shaped by human society.

The next chapter examines how human beings created God in the image of father, king, judge, and protector.

Sources

Boyer, Pascal. Religion Explained: The Evolutionary Origins of Religious Thought. Basic Books, 2001.

Eliade, Mircea. Shamanism: Archaic Techniques of Ecstasy. Princeton University Press, 1964.

Sacks, Oliver. Hallucinations. Knopf, 2012.

Tylor, Edward B. Primitive Culture. 1871.

Wallace, Anthony F. C. Religion: An Anthropological View. Random House, 1966.

Chapter 6: God as Father, King, Judge, and Protector

When the Universe Began to Look Human

Once human beings imagined invisible agents, the next step was almost natural.

They gave those agents human qualities.

The gods did not remain vague forces. They began to think, desire, command, love, punish, protect, bless, choose, remember, and become angry. They had preferences. They had rules. They had favorites. They could be pleased or offended. They could reward obedience and punish disobedience.

In other words, gods became persons.

This is one of the most revealing facts about religion. Human beings did not usually imagine God as a silent mathematical principle, a neutral physical law, or an impersonal force without emotion. They imagined God through relationships they already understood: father, mother, tribe, chief, king, judge, warrior, shepherd, master, protector, and lawgiver.

The universe became a larger version of human society.

This does not mean every culture imagined God in exactly the same way. Some religions imagined many gods. Some imagined one supreme God. Some imagined spirits, ancestors, divine forces, or sacred powers. But again and again, religious imagination gave the invisible world a human structure.

If human beings lived under kings, heaven often had a king.

If human beings lived under fathers, God often became father.

If human beings feared punishment, God became judge.

If human beings needed safety, God became protector.

If human beings fought wars, God became warrior.

If human beings valued law, God became lawgiver.

If human beings needed forgiveness, God became merciful.

This pattern matters.

It suggests that God was not simply found outside human life. God was shaped from within human life. The divine was built from human relationships, human fears, human authority, and human longing.

God became the face humanity placed upon the unknown.

The Father Above the World

One of the most powerful images of God is father.

This image carries deep emotional force because the father figure, in many cultures, represents authority, protection, discipline, identity, inheritance, and moral instruction. A father may be loved, feared, obeyed, admired, resisted, or longed for. That emotional range makes the father-image powerful religious material.

A child enters the world helpless. The child cannot feed himself, defend herself, understand danger, or explain pain. The child looks upward to adults for protection and explanation. A parent seems strong, knowing, and powerful. A parent can comfort, punish, rescue, approve, or reject.

To a small child, the parent can seem almost godlike.

Later, as adults, human beings may carry that emotional pattern into religion. The universe is large. Life is uncertain. Death is frightening. Suffering feels unfair. Human beings still want a protective presence above them. They want someone to hear them. They want someone to care. They want someone to say, "You are not alone."

God the Father answers that longing.

He watches. He knows. He protects. He commands. He forgives. He disciplines. He is above, yet personally concerned. He sees the hidden heart. He listens when no one else listens. He gives the believer a cosmic parent.

This is emotionally powerful.

But it also raises an important question: did human beings discover a heavenly father, or did they enlarge the father they already knew?

The image of God as father may reveal more about human emotional need than about divine reality. A frightened child wants a parent. A frightened adult may want one too. The longing is understandable. The belief built around that longing still needs evidence.

A lonely person may long for companionship. A grieving person may long for reunion. A frightened person may long for protection. A mortal person may long for immortality. In each case, the longing is real. The question is whether the object of the longing exists outside the mind.

God as father may be one of humanity's most tender inventions.

Tender, but still human.

Heaven Took the Shape of a Throne

In many religious traditions, God is not only father.

God is king.

This image comes from human society. For much of history, most people lived under chiefs, kings, emperors, lords, tribal elders, or rulers. Power was vertical. Commands came from above. Obedience was expected below. Law flowed downward. Disobedience brought punishment.

It is not surprising that heaven was imagined in similar terms.

God sits on a throne.

God rules.

God commands.

God has servants.

God has messengers.

God has enemies.

God has a kingdom.

God gives law.

God demands loyalty.

God punishes rebellion.

The earthly king became a model for the heavenly king. Sometimes the relationship worked in both directions. Kings claimed divine approval, and gods were imagined with royal qualities. Religion strengthened political power, and political power strengthened religion.

If the king ruled by God's will, disobedience to the king could become disobedience to God. If religious authorities blessed the ruler, power became sacred. If law was said to come from heaven, questioning law became dangerous.

This is one of the ways religion became woven into power.

The image of God as king gave the universe order. It told people that reality was not chaotic. Someone ruled. Someone commanded. Someone saw. Someone judged. In uncertain societies, this could be comforting.

But it could also support obedience.

A royal God can inspire humility, but also submission. He can inspire justice, but also hierarchy. He can comfort the powerless, but also strengthen the powerful. He can give moral order, but also protect unjust authority when rulers claim his name.

Again, the human origin is visible. Heaven often looks like the society that imagines it. A people governed by kings easily imagines a king in heaven. A people governed by law imagines a divine lawgiver. A people shaped by empire imagines cosmic empire.

If God were beyond all human culture, we might expect the divine image to escape human political forms. Yet religious history repeatedly shows God wearing the clothing of human power.

That is not a small clue.

The Judge No One Could Escape

God as judge may be one of the most psychologically powerful religious ideas.

Human society needs rules. People can lie, steal, betray, exploit, injure, and kill. Every group must find some way to control behavior. But human control is limited. A person may commit wrong in secret. A person may appear innocent while hiding guilt. A person may escape punishment. A powerful person may avoid justice. A clever person may deceive the community.

An all-seeing God solves this problem.

God sees what humans cannot see.

God knows hidden thoughts.

God knows secret actions.

God knows desire, intention, guilt, and motive.

God cannot be bribed.

God cannot be deceived.

God does not forget.

This idea gives moral weight to invisible behavior. Even when no one is watching, God is watching. Even when no court can prove wrongdoing, God knows. Even when the victim is powerless, God can judge. Even when the wicked prosper, God can punish later.

For societies struggling to maintain order, this belief is powerful. It may encourage cooperation. It may reduce cheating. It may make people feel accountable beyond human surveillance. It may comfort victims who never receive justice in this life.

But God as judge also creates fear.

If God sees everything, privacy disappears. Thoughts become crimes. Doubts become sins. Desires become moral failures. Natural feelings can become guilt. Human weakness can become spiritual danger. A person may feel watched not only in action, but in imagination.

This can shape the mind deeply.

A child taught that God sees every thought may develop obedience, but also fear. A believer may feel comforted by divine attention, but also burdened by constant self-judgment. Religious morality can inspire compassion, but it can also produce shame, anxiety, and submission.

The idea of God as judge is understandable because human beings care about justice. We want the universe to care about right and wrong. We want cruelty to be seen. We want hidden evil to be exposed. We want good people to matter.

God the Judge answers that longing.

But once again, need is not evidence.

The fact that we want perfect justice does not prove that a perfect judge exists.

The Protector We Wanted in the Dark

Human life has always been dangerous.

Children are fragile. Illness comes suddenly. Violence can erupt. Food can vanish. Animals can attack. Enemies can invade. Floods, storms, fires, droughts, and accidents can destroy years of effort in a day.

In such a world, the longing for protection is intense.

God the protector answers that longing.

He becomes shelter in danger, strength in weakness, light in darkness, comfort in fear, companion in loneliness, and hope in uncertainty. The believer can say, "Someone is with me. Someone sees me. Someone guards me. Someone has a plan."

This belief can give real emotional strength. People facing war, illness, poverty, or grief may endure because they believe they are held by a power greater than themselves. We should not dismiss this lightly. Human beings often survive unbearable situations by holding on to hope.

But there is a difficult problem.

Protection is uneven.

One child recovers, another dies.

One family escapes the flood, another is swept away.

One soldier survives, another does not.

One patient thanks God for healing, while another equally devout patient declines.

One person survives an accident and calls it a miracle, while the person beside him is killed.

If God is protector, why is protection so selective?

Believers often answer that God's ways are mysterious. But mystery can explain anything and therefore explains very little. If survival proves divine protection, what does death prove? If recovery proves God's love, what does suffering prove? If answered prayer proves God listens, what do unanswered prayers show?

A belief should not count only the comforting cases.

This issue will be examined more deeply in later books, especially The Failure of the Personal God. Here, the important point is that God as protector grew from a real human need. People wanted safety in a dangerous world. They wanted a guardian stronger than disease, death, weather, violence, and chance.

The longing is understandable.

But longing for protection does not prove that a cosmic protector exists.

The Warrior Behind the Tribe

Many gods were not gentle.

They fought. They conquered. They defended one people and destroyed another. They blessed armies, demanded loyalty, ordered punishment, and stood behind tribal victory. In many cultures, God did not belong equally to everyone. God belonged to us.

This is another sign of human origin.

Human beings live in groups. For most of history, group survival mattered more than universal love. A tribe needed loyalty. A city needed protection. A kingdom needed victory. A people surrounded by enemies needed courage. A ruler needed unity. Religion could provide all of these.

A warrior God turns battle into sacred duty.

The enemy is not merely another human group. The enemy becomes enemy of God. Victory becomes divine favor. Defeat becomes divine punishment. Land becomes promised. War becomes holy. Obedience becomes sacred. Dying for the group becomes spiritually meaningful.

This is one of religion's most dangerous powers.

When ordinary human conflict becomes divine conflict, compromise becomes harder. If God is on our side, the opponent is not merely wrong. The opponent is evil, impure, cursed, or rejected by heaven. Violence becomes easier when it is sanctified.

This does not mean religion always causes war. Human beings fight for land, food, power, fear, pride, revenge, and survival. But religion can intensify conflict by giving it cosmic meaning.

The warrior God reveals the tribal roots of religion. He is not the God of universal compassion. He is the God of boundary, battle, loyalty, and identity.

A truly universal divine truth should rise above tribe. Yet history shows gods repeatedly tied to tribe, land, nation, empire, and conquest.

This suggests that gods often served human groups before they served universal morality.

The Lawgiver Above Question

Human communities need rules. Without rules, trust collapses. People must know what is permitted, forbidden, rewarded, and punished. They must regulate marriage, property, inheritance, food, conflict, sexuality, violence, debt, ritual, and authority.

Religion gave rules divine weight.

A rule made by humans can be questioned. A rule said to come from God becomes harder to challenge. It stands above debate. It becomes sacred. It carries fear.

This gave religion great social power.

Do not steal, because God commands it.

Do not kill, because God forbids it.

Do not eat this food, because God has made it unclean.

Do not marry that person, because God disapproves.

Do not question the priest, because he speaks for God.

Do not disobey the king, because he rules by divine will.

Sacred law can support social order. Some religious laws may protect community life, reduce violence, encourage honesty, and support charity. But sacred law can also freeze ancient customs and make reform difficult. Rules born in one time and place may be treated as eternal. Human prejudice can become divine command. Social hierarchy can become sacred duty.

This is why the idea of God as lawgiver must be examined carefully.

When a society says, "God commands this," we must ask: how do we know? Who speaks for God? Who benefits from this command? Does the rule reduce suffering or increase it? Does it treat people with dignity? Can it be corrected if it is wrong?

A human law can be revised. A divine law is often defended even when it harms.

This does not prove that all religious law is bad. It proves that calling a law divine can protect it from moral examination.

The lawgiver God may have helped societies organize behavior. But he also gave human authorities a powerful tool: the ability to place their rules beyond question.

God as the Mirror of Society

When we compare religions, a pattern appears.

Gods often mirror the societies that worship them.

In monarchies, God is often king.

In patriarchal societies, God is often father.

In warrior cultures, God often fights.

In legalistic cultures, God often judges.

In agricultural societies, gods often control fertility, rain, and harvest.

In tribal societies, gods often protect the group.

In societies with strict purity codes, gods often care deeply about food, sex, blood, and ritual cleanliness.

This pattern does not look accidental.

It suggests that human beings project their own social world onto the cosmos. They take familiar forms of power and enlarge them. They take human authority and place it in heaven. They take the father, king, judge, warrior, and lawgiver, then make them divine.

This was Ludwig Feuerbach's great insight: theology often reveals anthropology. What people say about God may reveal what they value, fear, admire, and need as human beings.

If a society imagines God as wrathful, we learn something about that society.

If it imagines God as tender, we learn something.

If it imagines God as obsessed with purity, we learn something.

If it imagines God as a warrior, we learn something.

If it imagines God as a judge, we learn something.

The study of God becomes the study of humanity.

This does not settle every philosophical question about God. But it strongly suggests that many God-images are human creations. They carry human fingerprints. They change across history. They adapt to culture. They reflect social structure.

Gods look less like independent discoveries and more like portraits painted by human need.

The Comfort and Danger of a Human-Like God

A human-like God is emotionally powerful because human beings understand persons.

A law of physics cannot love you.

Gravity cannot forgive you.

Evolution cannot hear your prayer.

The universe cannot hold your hand.

But a personal God can.

That is why a human-like God is so attractive. People want not only explanation, but relationship. They want to be known. They want to be forgiven. They want to be protected. They want their suffering to be seen. They want their life to matter to someone larger than death.

A personal God gives the universe a heart.

This is the comfort.

But there is also danger.

A human-like God can carry human anger, jealousy, favoritism, violence, and prejudice. He can be used to bless one tribe and curse another. He can be used to control women, children, outsiders, doubters, and the poor. He can be used to make human authority sacred. He can turn disagreement into sin. He can turn curiosity into rebellion. He can turn doubt into guilt.

When God becomes too human, human flaws can become divine.

This is why the image of God must be examined. If God looks like a father, king, judge, warrior, and lawgiver, we must ask whether we are seeing reality or seeing ourselves enlarged.

Perhaps God made humans in his image.

But perhaps humans made God in theirs.

That question stands at the center of this book.

Chapter Summary

Human beings did not usually imagine God as a neutral force or abstract principle. They imagined God in human terms: father, king, judge, protector, warrior, and lawgiver. These images came from human society and human need. The child's dependence on parents helped shape God as father. Political authority helped shape God as king. The desire for justice helped shape God as judge. Fear and danger helped shape God as protector. Tribal conflict helped shape God as warrior. Social order helped shape God as lawgiver.

These images made God emotionally powerful because they turned the universe into a relationship. People could love God, fear God, obey God, ask God for help, and imagine God watching over them. But these images also reveal the human roots of religion. Gods often mirror the societies that create them. They carry the structure of family, tribe, law, monarchy, war, and authority.

God was not only an answer to nature.

God was also an enlargement of human society.

Key Points

Human beings often imagined God in familiar human roles.

God as father answered the longing for protection, love, and guidance.

God as king reflected human political power and hierarchy.

God as judge answered the desire for moral order and hidden justice.

God as protector grew from fear and helplessness in a dangerous world.

Warrior gods often reflected tribal identity and conflict.

God as lawgiver gave human rules sacred authority.

Gods often mirror the societies that worship them.

A human-like God comforts people, but can also sanctify human power, fear, and prejudice.

Transition to the Next Chapter

Once human beings imagined gods as powerful invisible persons, they naturally tried to influence them.

If the gods could be pleased, people would offer gifts. If they could be angered, people would seek forgiveness. If they could protect, people would pray. If they could punish, people would obey. If they could control rain, harvest, fertility, war, and disease, people would perform rituals to win their favor.

The next chapter examines rituals, prayer, sacrifice, and the deep human need to feel some control over an uncertain world.

Sources

Armstrong, Karen. A History of God: The 4,000-Year Quest of Judaism, Christianity and Islam. Knopf, 1993.

Feuerbach, Ludwig. The Essence of Christianity. 1841.

Freud, Sigmund. The Future of an Illusion. 1927.

Hume, David. The Natural History of Religion. 1757.

Norenzayan, Ara. Big Gods: How Religion Transformed Cooperation and Conflict. Princeton University Press, 2013.

Chapter 7: Rituals, Prayer, Sacrifice, and the Illusion of Control

When Helplessness Needed an Action

Human beings do not tolerate helplessness easily.

When a child is sick, the parent wants to act. When the sky gives no rain, the farmer wants to act. When the enemy approaches, the tribe wants to act. When death enters the home, the family wants to act. When fear fills the body and no clear solution is available, doing nothing feels unbearable.

This may be one of the deepest roots of ritual.

Ritual gives action to helplessness.

It gives the frightened mind a sequence.

Light the fire.

Say the words.

Bring the offering.

Wash the body.

Bow the head.

Touch the sacred object.

Repeat the prayer.

Follow the rule.

Wait for mercy.

Even when the ritual does not change nature, it changes the person performing it. It gives fear a shape. It gives anxiety a movement. It gives the group something shared to do. It tells people they are not simply standing before disaster with empty hands.

This is why ritual became so powerful in religion. It gave human beings a way to respond to forces they could not understand and could not control.

If rain did not come, they could pray.

If disease spread, they could purify.

If a child was born, they could bless.

If someone died, they could bury, chant, burn incense, or speak sacred words.

If the tribe feared war, they could sacrifice.

If guilt became heavy, they could confess.

Religion did not only tell people what to believe.

It told them what to do.

That was its great power.

A belief may live in the mind, but a ritual enters the body. It uses voice, hands, posture, rhythm, food, water, fire, blood, music, movement, silence, and repetition. It makes the invisible feel physical. It makes the sacred feel close.

Ritual turns an idea into an experience.

And experience can feel more convincing than argument.

Prayer Made the Universe Personal

Prayer is one of the most intimate acts in religion.

A person may pray in a temple, church, mosque, shrine, field, hospital room, prison cell, battlefield, bedroom, or silent corner of the heart. Prayer may be spoken aloud, whispered, sung, memorized, improvised, or held silently inside the mind.

Through prayer, the invisible becomes personal.

The believer is no longer facing the universe alone. Someone is listening. Someone knows. Someone may help. Someone may forgive. Someone may heal. Someone may protect. Someone may guide.

This is emotionally powerful because human beings are social creatures. We are built to speak, ask, confess, plead, thank, and hope in relationship with others. Prayer takes these human habits and directs them toward an unseen listener.

A frightened person prays for protection.

A grieving person prays for reunion.

A sick person prays for healing.

A guilty person prays for forgiveness.

A lonely person prays to be heard.

A poor person prays for justice.

A dying person prays for peace.

Prayer can calm the mind. It can organize emotion. It can reduce loneliness. It can help people endure pain. It can create courage when no human help seems near. For these reasons, prayer can have real psychological value.

But psychological value is not supernatural evidence.

A person may feel comforted by prayer because prayer creates focus, hope, emotional release, rhythm, and the feeling of relationship. The comfort may be real even if no divine being is answering.

A grieving spouse may be comforted by speaking to the dead.

A lonely person may be comforted by speaking aloud to an imagined listener.

A frightened patient may feel stronger after prayer because prayer has gathered scattered fear into words.

The emotional effect can be genuine without proving that the listener exists.

This distinction matters.

Prayer may reveal the depth of human need. It may reveal longing, gratitude, guilt, fear, and hope. It may reveal the mind's ability to create a felt relationship with an unseen presence.

But prayer does not automatically prove God.

It may prove that human beings need to be heard.

Sacrifice Was Bargaining With Power

Sacrifice is one of the oldest and most dramatic forms of religious behavior.

Human beings gave food, animals, crops, blood, wealth, time, comfort, sexuality, obedience, and sometimes even human life to invisible powers. The logic was simple and deeply human: if the gods can give and take, perhaps they can be persuaded.

Sacrifice was bargaining with power.

Take this goat, and give us rain.

Take this grain, and protect our children.

Take this blood, and forgive our guilt.

Take this offering, and spare the village.

Take this life, and save the rest.

To modern readers, sacrifice may seem strange or cruel. But emotionally, it follows a familiar human pattern. We give gifts to people whose favor we want. We apologize when we offend. We bring offerings to leaders. We try to repair damaged relationships. We bargain when afraid.

Early religion extended these human behaviors toward invisible beings.

If gods were like kings, they might accept tribute.

If gods were like parents, they might forgive after apology.

If gods were like judges, they might reduce punishment.

If gods were like warriors, they might protect loyal followers.

Sacrifice made the relationship with the divine feel practical. It gave human beings a way to influence unseen power. It also created seriousness. A prayer may be only words, but sacrifice costs something. The more precious the offering, the more intense the act feels.

This is one reason sacrifice became central in many religions. It placed pain, loss, and devotion into visible form. It told the group: we are serious. We are willing to give something important. We are trying to restore order.

But sacrifice also reveals the human construction of religion. The gods are imagined as beings who can be pleased, bribed, fed, honored, soothed, or satisfied. These are human social patterns projected onto invisible power.

A scientific view asks a simple question: did sacrifice bring rain, heal disease, stop earthquakes, or change the motion of planets?

There is no reliable evidence that it did.

But sacrifice did something else. It changed the human group. It created unity, fear, obedience, memory, and authority. It made belief visible.

That may explain why sacrifice survived for so long.

Ritual Turned Panic Into Order

Fear is chaotic.

The body trembles. The mind races. Thoughts scatter. Time feels distorted. People become desperate, angry, submissive, or confused. Ritual gives fear an order it does not naturally have.

This is one reason rituals are so common around birth, puberty, marriage, sickness, death, planting, harvest, war, and disaster. These are moments when life feels unstable. Ritual creates structure at the edge of uncertainty.

At birth, ritual welcomes the child.

At puberty, ritual marks the transition.

At marriage, ritual creates public commitment.

At sickness, ritual seeks healing or comfort.

At death, ritual helps the living face loss.

At harvest, ritual expresses gratitude.

At war, ritual creates courage.

At disaster, ritual restores group identity.

Rituals say: we know what to do here.

That sentence is powerful. Human beings suffer more when they do not know what to do. Ritual gives a script. It tells people where to stand, what to say, when to be silent, what to wear, how to move, whom to follow, and what the event means.

This does not make the ritual scientifically true. It makes it socially and emotionally effective.

A funeral ritual may not prove life after death, but it helps the living grieve. A healing ritual may not remove infection, but it may reduce panic. A harvest ritual may not control weather, but it may strengthen community. A purification ritual may not remove sin, but it may relieve guilt.

Ritual often works on the human mind, not on nature.

This is a crucial distinction.

Religion may claim that ritual changes the world. In many cases, the stronger evidence is that ritual changes the people who perform it.

It reduces fear.

It creates belonging.

It marks meaning.

It strengthens memory.

It gives authority.

It makes invisible belief visible.

In that sense, ritual is not useless. It is humanly powerful. But its power does not prove the supernatural claims attached to it.

Repetition Made Belief Feel True

A ritual repeated once may be memorable. A ritual repeated for generations becomes tradition. A ritual repeated from childhood becomes identity.

Repetition is one of the great forces behind religious belief.

A child hears the same prayer every night. The same festival returns every year. The same sacred story is told by parents, grandparents, priests, teachers, and neighbors. The same gestures are repeated: bowing, kneeling, touching, washing, lighting, singing, fasting, offering.

Long before the child understands theology, the body has already learned the rhythm of belief.

Over time, repetition creates familiarity.

Familiarity can feel like truth.

This happens in many areas of life. A story heard many times feels more believable. A phrase repeated by many people feels more authoritative. A custom practiced for centuries feels sacred. A belief held by loved ones feels morally safe.

Religion uses repetition better than almost any other human system.

It places belief into time: daily prayers, weekly worship, annual festivals, life-cycle rituals, birth ceremonies, coming-of-age rituals, weddings, funerals, holy days, fasting periods, and pilgrimage seasons. The believer does not encounter religion only as an idea. The believer lives inside its calendar.

This gives religion extraordinary durability.

A philosophical argument may be forgotten. A scientific fact may be misunderstood. But a ritual repeated with music, family, food, dress, memory, fear, hope, and community becomes part of the person.

To question it later may feel like questioning one's own life.

This is why inherited belief can be so strong. It is not merely accepted. It is practiced into the body.

The hands learn it.

The voice learns it.

The knees learn it.

The stomach learns it through fasting and feasting.

The eyes learn it through symbols.

The ears learn it through songs.

The heart learns it through family.

By the time reason asks questions, belief may already feel ancient inside the self.

The Crowd That Enlarged Faith

Ritual is rarely only private. It is often communal.

A group gathers. They sing the same words. They move together. They bow together. They listen together. They mourn together. They eat together. They repeat the same sacred story. They feel themselves becoming one body.

This can be deeply moving.

Human beings are social animals. We are affected by the emotions of others. A crowd can intensify fear, joy, anger, courage, grief, or devotion. Music can synchronize breathing. Chanting can synchronize voice. Movement can synchronize bodies. Shared silence can create deep emotional unity.

Religion often uses this beautifully.

A lone belief may be fragile. A shared belief becomes powerful. When everyone around us believes, belief feels normal. When everyone prays, prayer feels natural. When everyone kneels, kneeling feels right. When everyone cries during a ritual, emotion confirms belonging.

The group becomes evidence.

A person thinks: so many people cannot be wrong. My parents believed. My grandparents believed. My people have believed for generations. How could all of this be merely human?

But the number of believers does not decide truth.

Large groups can share false beliefs. Entire societies have believed in omens, witchcraft, astrology, racial hierarchy, sacred kings, divine wars, and many gods now abandoned. Social agreement makes belief stronger, but not necessarily truer.

The community that prays together may become more united. That is real. But unity is not evidence that the prayer reaches heaven.

This distinction is vital.

Religion often proves its social power by its emotional effects on the group. But social power does not prove supernatural truth. It proves that human beings can be deeply shaped by shared ritual and shared belief.

Purity Gave Fear a Rulebook

Many religions developed rules of purity and impurity.

Certain foods became forbidden. Certain animals became unclean. Certain bodily fluids became dangerous. Certain sexual acts became sinful. Certain places required washing. Certain people were kept apart. Certain objects could not be touched. Certain days were holy. Certain words could not be spoken.

Purity rules are powerful because they turn invisible danger into visible practice.

Wash this way.

Eat this, not that.

Touch this, avoid that.

Enter here, not there.

Wear this.

Do not mix these things.

Do not cross this boundary.

Rules reduce uncertainty. They give people a sense that danger can be managed. If disease, misfortune, guilt, or divine anger feels invisible, purity laws make the invisible manageable through behavior.

Some purity rules may have had practical value in certain environments. Food practices, hygiene habits, and social boundaries can sometimes reduce risk or preserve group identity. But many purity rules also reflect fear, taboo, hierarchy, and control.

The idea of impurity can become cruel when applied to people. Women, lower classes, foreigners, sexual minorities, the sick, the disabled, mourners, and outsiders have often been labeled impure in various religious systems. Once a person is called impure by God, discrimination becomes sacred.

This is one danger of religious rule-making. Human fears become divine commands.

Purity rules also show how religion connects body and morality. Eating, sex, menstruation, birth, death, illness, and touch become spiritually charged. The body becomes a battlefield between sacred and polluted.

Modern science gives a different approach. Hygiene matters, but not because of ritual impurity. Food safety matters, but not because an animal is spiritually cursed. Disease prevention matters, but not because the sick are morally unclean. Public health replaces taboo with evidence.

Science asks what actually causes harm.

Religion often asks what violates sacred order.

Those are not the same question.

The Priest Who Knew the Correct Words

Ritual creates power because someone must know how to perform it correctly.

What words should be spoken?

What offering is acceptable?

What day is holy?

What food is forbidden?

What action removes guilt?

What prayer reaches God?

What sacrifice restores favor?

What ceremony protects the dead?

What rule purifies the sinner?

The person who claims to know these answers becomes important.

This is one path by which priests, shamans, ritual experts, elders, and religious institutions gained authority. If the invisible world matters, then the people who interpret the invisible world hold power. They can bless, forbid, purify, condemn, forgive, marry, bury, exclude, or welcome. They can tell people what God wants.

This does not mean all religious leaders were dishonest. Many were sincere. Many served their communities. Many comforted the grieving and helped preserve social order. But sincerity does not remove power.

Ritual authority can become social authority.

The expert in sacrifice may influence the farmer.

The expert in prayer may influence the sick.

The expert in purity may influence women and families.

The expert in scripture may influence rulers.

The expert in burial may influence the grieving.

The expert in sin may influence everyone.

Once religious authority becomes established, questioning it can become dangerous. A person who doubts the ritual may be accused of doubting God. A person who questions the priest may be accused of rebellion. A person who asks for evidence may be accused of arrogance.

This is how human authority can hide behind divine authority.

Ritual may begin as comfort, but it can become control.

Comfort Is Real, Control Is Dangerous

Ritual and prayer can comfort people. This should be acknowledged honestly.

A person praying beside a hospital bed may become calmer.

A family performing a funeral ritual may feel supported.

A community singing together may feel united.

A person confessing guilt may feel relieved.

A pilgrim walking toward a sacred place may feel renewed.

These experiences are real. They should not be mocked.

But we must ask what they prove.

They prove that human beings respond deeply to symbols, rhythm, memory, community, words, music, and action. They prove that ritual can shape emotion. They prove that prayer can organize fear and hope. They prove that shared belief can strengthen belonging.

They do not prove that a supernatural being is controlling the outcome.

This distinction is not cold. It is honest.

In medicine, a patient may feel better after reassurance. That matters. Reassurance can reduce fear. A calm patient may sleep better, cooperate with treatment, and endure suffering more bravely. But reassurance does not replace antibiotics when infection is present. It does not replace insulin when glucose is dangerously high. It does not replace surgery when the appendix has ruptured.

Comfort has value, but comfort is not cure.

In the same way, prayer may comfort, but comfort is not evidence of divine intervention. Ritual may reduce anxiety, but anxiety relief is not proof that gods were pleased. Sacrifice may unite a group, but unity is not proof that invisible beings accepted the offering.

The human effect is real.

The supernatural claim remains unproven.

That is one of the central lessons of this chapter.

From Ritual to Reason

Ritual was one of humanity's early answers to helplessness. Science became a later and better answer.

When people did not understand disease, they prayed and purified. When they understood microbes, they washed hands, sterilized instruments, purified water, developed vaccines, and used antibiotics.

When people did not understand drought, they performed rain rituals. When they understood agriculture and climate better, they built irrigation, studied weather, stored grain, improved soil, and developed drought-resistant crops.

When people did not understand eclipses, they feared omens. When they understood astronomy, they predicted eclipses with precision.

When people did not understand childbirth, they used charms and prayers. When they understood anatomy, infection, bleeding, nutrition, and obstetric care, maternal survival improved.

This is not an argument against comfort, community, or meaning. It is an argument against mistaking ritual for explanation.

Ritual asks: what sacred action must we perform?

Science asks: what is actually happening?

Ritual asks: who have we offended?

Science asks: what causes this event?

Ritual asks: how can we please the invisible power?

Science asks: what evidence can guide effective action?

The difference matters because reality responds to causes, not merely to feelings. A fever does not fall because we understand it emotionally. A bridge does not stand because we bless it. A crop does not grow because we perform the correct dance. A storm does not turn aside because we chant.

Human beings needed more than ritual.

They needed knowledge.

That movement from ritual to reason is one of the greatest stories of human progress. It does not remove every sorrow. It does not solve every mystery. But it gives us a better way to reduce suffering.

And that is why God slowly became less necessary as an explanation.

Chapter Summary

Ritual, prayer, and sacrifice arose from the deep human need to act in the face of helplessness. When people did not understand disease, weather, death, fertility, disaster, or chance, ritual gave them something to do. Prayer made the invisible personal. Sacrifice turned fear into bargaining. Purity rules gave invisible danger a visible rulebook. Communal ritual created belonging and emotional unity.

These practices were powerful because they changed human feeling and strengthened group identity. They reduced anxiety, gave structure to grief, and made uncertainty more bearable. But psychological comfort does not prove supernatural truth. Ritual may change the human mind without changing nature. Prayer may comfort without proving that anyone is listening. Sacrifice may unite a community without influencing the sky, disease, or death.

Religion often used ritual to create both comfort and authority. Science later offered a different path: not appeasement of invisible powers, but investigation of natural causes.

Ritual gave humanity something to do.

Science gave humanity something better to know.

Key Points

Ritual gives action to helplessness.

Prayer makes the invisible feel personal and emotionally present.

Sacrifice reflects bargaining with imagined divine power.

Ritual can reduce fear, organize grief, and strengthen community.

Repetition makes belief feel familiar and true.

Communal worship can intensify emotion and belonging.

Purity rules turn fear into behavior, but can also create control and exclusion.

Ritual authority can become social and political authority.

Comfort is real, but comfort is not evidence of supernatural truth.

Science replaced ritual appeasement with investigation, prevention, and effective action.

Transition to the Next Chapter

Ritual did more than comfort individuals. It joined people together.

A person may believe alone, but religion becomes powerful when belief becomes community. Shared rituals become shared identity. Shared stories become tribal memory. Shared gods become boundaries between us and them.

The next chapter examines religion as tribe, identity, and belonging, and why belief became one of the strongest forces holding human groups together.

Sources

Bellah, Robert N. Religion in Human Evolution: From the Paleolithic to the Axial Age. Harvard University Press, 2011.

Durkheim, Émile. The Elementary Forms of Religious Life. 1912.

Malinowski, Bronislaw. Magic, Science and Religion and Other Essays. 1948.

Turner, Victor. The Ritual Process: Structure and Anti-Structure. Aldine, 1969.

Whitehouse, Harvey. Modes of Religiosity: A Cognitive Theory of Religious Transmission. AltaMira Press, 2004.

Chapter 8: Religion as Tribe, Identity, and Belonging

The Belief That Became a Home

Religion is rarely only a belief in the mind.

For many people, religion is home.

It is the sound of a prayer heard in childhood. It is a festival meal. It is the smell of incense, candles, flowers, old books, or familiar food. It is a grandmother telling a sacred story. It is a father's instruction. It is a mother's song before sleep. It is the wedding blessing, the funeral chant, the newborn's name, the holy day that returns each year, and the familiar words repeated when life becomes uncertain.

This is why religion becomes so powerful.

A person does not usually receive religion as a bare idea. A person receives it as family, language, belonging, fear, music, food, marriage, grief, hope, and identity. The belief is wrapped inside life itself.

That makes religion difficult to question.

When a person begins to doubt God, the doubt may not feel like a simple intellectual question. It may feel like betrayal. Betrayal of parents. Betrayal of ancestors. Betrayal of community. Betrayal of childhood. Betrayal of the dead. Betrayal of the people who loved, fed, taught, protected, and raised that person.

This is one reason religious belief can survive even when evidence is weak. Religion is not defended only by argument. It is defended by attachment.

People do not easily abandon what gave them belonging.

This chapter examines religion as tribe, identity, and community. It asks how belief in gods became one of the strongest social bonds in human history. It also asks a difficult question: if religion binds people together, does that make it true?

The answer is no.

A belief can unite a tribe and still be false. A myth can hold a people together and still be human-made. A shared ritual can create deep emotional power and still not prove that heaven exists.

Religion became powerful because it did not only explain the world.

It made people feel that they belonged inside it.

The Sacred Circle of Us

Every human group needs a sense of "us."

Who are we?

Who are our people?

What story do we share?

What rules bind us?

What do we celebrate?

What do we fear?

Who belongs inside the circle?

Who stands outside?

Religion answered these questions with unusual force.

A group that shares a god becomes more than a collection of individuals. It becomes a sacred community. Its land may become holy. Its ancestors may become chosen. Its laws may become divine. Its suffering may become meaningful. Its enemies may become enemies of God. Its survival may feel like part of a cosmic story.

This gives identity tremendous strength.

People may sacrifice for a group when they believe the group is sacred. They may endure hardship, obey rules, defend boundaries, preserve stories, and teach traditions to children. Religion gives the group a memory older than any living person. It tells each generation, "You are part of something larger than yourself."

This can be beautiful.

A sacred community can support the poor, comfort the grieving, teach children, preserve discipline, and give dignity to people whose lives are otherwise hard. It can offer warmth in a lonely world.

But the same sacred circle can also exclude.

If there is an "us," there is often a "them."

Those outside the circle may be seen as impure, lost, dangerous, sinful, cursed, inferior, or chosen for conversion. Religion can create compassion within the group while hardening suspicion toward those outside it.

This is one of religion's double powers.

It can make strangers into brothers.

It can also make neighbors into enemies.

The Story That Held the Tribe Together

Every group needs a story.

A nation has a story. A family has a story. A political movement has a story. A religion has a story. These stories tell people where they came from, what they suffered, who their heroes were, what their enemies did, what God wants, and what future awaits them.

A shared story helps people live together.

It gives the child a place inside history. It gives the old a reason to preserve memory. It gives the group a moral map. It tells people what is honorable, shameful, sacred, forbidden, and worth defending.

Religious stories are especially powerful because they are not presented as ordinary stories. They are sacred stories. They are often said to come from God, prophets, sages, revelations, ancestors, or holy events. That gives them authority beyond ordinary memory.

A community may say:

We were chosen.

We were rescued.

We were tested.

We were given law.

We were promised land.

We were protected by God.

We were betrayed by outsiders.

We must remain pure.

We must teach our children.

Such stories can give people courage. They can preserve a group through exile, poverty, persecution, migration, humiliation, or war. They can keep identity alive across centuries.

But sacred stories also create a problem. Once a story is treated as divine, it becomes hard to correct. It can survive even when history complicates it. It can be repeated even when it contains cruelty. It can justify conquest, exclusion, obedience, or fear. It can make one group feel more loved by God than another.

A story can give meaning without being literally true.

A story can preserve identity without proving divine revelation.

A story can be emotionally necessary for a community and still be a human creation.

The power of sacred stories tells us much about human beings. It tells us that we need meaning, memory, and belonging. It does not prove that the story came from God.

Belonging Can Be Stronger Than Evidence

Many people think beliefs change mainly because of evidence.

Sometimes they do. But often, belonging is stronger than evidence.

A person may continue to believe because everyone he loves believes. A woman may stay in a religion because leaving would break her family. A young man may hide doubt because he fears losing community. An older person may avoid difficult questions because religion is tied to marriage, children, friends, festivals, and the memory of parents.

This is not cowardice. It is human attachment.

Belief is often social.

We believe with our people.

A person born in one community learns one set of sacred truths. A person born elsewhere learns another. Each community may say its own belief is obvious, holy, and true. But from the outside, we can see how strongly geography and family shape certainty.

A child born in a Muslim home is likely to hear Islamic prayers, stories, and moral rules. A child born in a Christian home is likely to hear Christian stories, hymns, and teachings. A child born in a Hindu home is likely to grow up with Hindu festivals, deities, rituals, and family practices. A child born in a secular home may grow up without fear of hell or expectation of divine intervention.

Each child receives a world.

Later, as an adult, each may feel that the inherited world is simply true.

This is why belief can be so resistant to evidence. Changing belief may mean losing social belonging. It may mean disappointing parents. It may mean being treated as fallen, arrogant, sinful, or corrupted. It may mean exile from the emotional world of childhood.

Evidence asks the mind to change.

Belonging asks the heart to stay.

Religion survives because it holds both the mind and the heart.

The Boundary Between Us and Them

Religion creates strong boundaries.

This is our God.

This is our scripture.

This is our food.

This is our ritual.

This is our holy day.

This is our way to marry.

This is our way to bury the dead.

This is our moral law.

This is who we are.

Boundaries can protect identity. They help groups survive. They preserve customs, language, memory, and values. A group without boundaries may disappear into larger cultures. Religion can act like a fence around a people.

But fences have two sides.

The same boundary that protects identity can create suspicion of outsiders. The outsider may be seen as unclean, dangerous, ignorant, immoral, cursed, lost, or spiritually inferior. Children may be warned not to marry outsiders, eat with outsiders, trust outsiders, or learn from outsiders. Doubt itself may be treated as contamination from outside influence.

This is how religion can make human difference feel absolute.

Ordinary cultural difference becomes sacred difference.

Different food becomes impurity.

Different worship becomes sin.

Different belief becomes rebellion.

Different identity becomes threat.

This process can be dangerous. When a group believes God stands inside its boundary, it may treat those outside the boundary with less compassion. Outsiders become easier to fear, convert, control, or condemn.

This does not mean religious people cannot love outsiders. Many do. Many religious traditions also teach kindness to strangers. But religion has a repeated historical tendency to make group boundaries sacred.

Once a boundary becomes sacred, crossing it becomes more than social change.

It becomes spiritual danger.

The Chosen People and the Unchosen World

Many religious groups have believed that they are specially chosen, specially guided, specially loved, or specially favored by God.

This belief is powerful. It gives dignity to suffering. It gives courage to small groups. It tells a poor or oppressed people that they are not forgotten. It gives the community a noble role in history. It says, "We matter because God has chosen us."

But the idea of being chosen has a shadow.

If one group is chosen, what about the others?

Are they less loved?

Less enlightened?

Less pure?

Less obedient?

Less human?

The chosen identity can easily become spiritual pride. It can make a group feel superior. It can turn history into a story where one people stands closer to God than all others. It can justify conquest, exclusion, conversion, or contempt.

This pattern is not limited to one religion. Many communities, in different ways, have imagined themselves as specially favored by divine power. They may not all use the word chosen, but the emotional structure is similar.

Our way is the true way.

Our scripture is the final word.

Our prophet is the last messenger.

Our land is sacred.

Our people have a divine mission.

Our tradition is pure.

This is understandable from a social point of view. Groups survive better when members believe the group matters. A sacred mission strengthens loyalty. It makes sacrifice meaningful. It gives ordinary life cosmic importance.

But usefulness is not truth.

The belief that a group is chosen may help that group endure. It may give courage. It may preserve identity. But it does not prove that the universe has selected that group for divine favor.

It may prove only that human groups need stories that make them feel important.

The Heavy Price of Leaving

For many people, leaving religion is not like changing an opinion.

It is like leaving a country.

A person may lose family trust. Friendships may weaken. Marriage may suffer. Parents may grieve as if the child has died spiritually. The community may gossip. The person may be called arrogant, immoral, rebellious, possessed, corrupted, or lost.

The cost can be heavy.

This social cost helps religion survive.

People may doubt privately but remain publicly loyal. They may stop believing in the heart but continue rituals for family peace. They may teach children what they themselves no longer fully accept because they fear conflict. They may attend worship to avoid questions. They may hide honest thoughts even from the people closest to them.

This means religious belief is not always as strong as it appears. Public conformity can be mistaken for private conviction.

But it also shows the power of belonging. Religion becomes part of the social body. Leaving it can feel like cutting oneself away from the people one loves.

This is one reason a humane atheist must speak carefully. Religious belief is not merely an error in logic. It is often tied to belonging, loyalty, memory, fear, and love. If we attack it crudely, people will defend not only the belief, but the whole emotional world around it.

A better way is to separate the person from the belief.

The believer is not the enemy.

The question is whether the belief is true.

A person can leave a belief without hating the people who gave it to him.

A person can question inherited religion without despising family.

A person can seek truth without becoming cruel.

This distinction matters deeply.

Sacred Loyalty Can Blind Good People

Loyalty can be beautiful.

Loyalty helps families care for each other. It helps communities survive hardship. It helps friends stand together. It helps people keep promises. It gives courage in difficult times.

But sacred loyalty can also create blindness.

When a belief is tied to group identity, people may defend it even when it harms others. They may excuse cruelty if it is done by their side. They may ignore contradictions in their own tradition while criticizing another. They may call their own violence justice and another group's violence evil. They may protect leaders because those leaders represent the faith.

This is not unique to religion. Political parties, nations, ethnic groups, and ideologies can do the same. But religion adds another layer: divine approval.

If our group belongs to God, then defending the group may feel like defending God.

If our scripture is perfect, then criticism may feel like blasphemy.

If our leader speaks for God, then questioning him may feel like rebellion.

If our tradition is sacred, then reform may feel like betrayal.

This can produce moral blindness. People may tolerate injustice because it is old. They may defend prejudice because it is traditional. They may obey cruel commands because they are called holy. They may silence victims to protect the reputation of the religious group.

The problem is not loyalty itself.

The problem is loyalty placed above truth and compassion.

A moral life requires the courage to ask whether our own group may be wrong.

Religion often teaches people to examine outsiders.

Science and reason ask us to examine ourselves.

That is much harder.

Community Without Chains

One of the reasons people fear life without religion is that they fear losing community.

This fear is real. Human beings need belonging. We need people who know us, remember us, help us, celebrate with us, grieve with us, and stand beside us when life becomes difficult.

Religion has often provided that. It has given meals, festivals, songs, weddings, funerals, charity, moral teaching, and connection across generations. It has helped lonely people feel seen. It has helped grieving people feel supported. It has helped the poor survive. It has given children a place inside a larger story.

These human goods should not be dismissed.

But the question is whether community requires belief in God.

It does not.

Human beings can build community around compassion, truth, art, science, service, education, friendship, justice, family, and mutual care. We can gather without pretending that supernatural claims are true. We can comfort the grieving without promising heaven. We can celebrate birth without invoking divine plan. We can mourn the dead while honoring the real life they lived. We can build moral communities without fear of hell.

The loss of religion does not have to mean the loss of belonging.

It may mean the difficult work of building more honest forms of belonging.

A community does not become less human because it is secular. It becomes less human only when it loses compassion.

This is important for the entire No God Required series. The goal is not to leave people isolated after questioning faith. The goal is to show that meaning, morality, and community can survive without God.

Indeed, they may become more honest when they no longer depend on claims that cannot be proven.

What Belonging Proves and What It Does Not

Religion's power as belonging is real.

It can create loyalty. It can preserve memory. It can comfort grief. It can turn strangers into family. It can give children identity. It can help people endure suffering. It can make life feel woven into something larger than the self.

But none of this proves God.

A family story can unite a family and still contain exaggerations.

A national myth can inspire courage and still distort history.

A political movement can create solidarity and still be wrong.

A religious tradition can build community and still make false supernatural claims.

Human beings are meaning-making animals. We create bonds through stories, songs, symbols, shared meals, common suffering, and repeated rituals. Religion is one of the strongest examples of this human power.

But social strength is not the same as truth.

This distinction allows us to be fair. We can acknowledge the emotional and social value of religion without accepting its supernatural claims. We can understand why people stay without pretending that staying proves the belief true. We can see the beauty of belonging while also recognizing the danger of sacred boundaries.

Religion answers the human need for home.

But home is not the same as evidence.

Chapter Summary

Religion became powerful not only because it explained nature or comforted fear, but because it created belonging. It gave people identity, shared stories, rituals, moral rules, festivals, boundaries, and a sense of being part of something larger than themselves. For many people, religion is not simply a belief. It is family, childhood, memory, community, and home.

This helps explain why religious belief is so durable. People often defend religion not only because they think it is true, but because it connects them to the people and traditions they love. Religion creates a sacred circle of "us," but that circle can also create suspicion or rejection of "them." It can inspire care within the group while hardening boundaries against outsiders.

Religion's social power is real. It can comfort, organize, unite, and preserve identity. But social power does not prove supernatural truth. A belief can bind a community and still be false. A sacred story can give meaning and still be human-made.

Religion became powerful because it became home.

But a home can be loved without being a map of reality.

Key Points

Religion often functions as identity, not merely belief.

Shared stories, rituals, and symbols help create strong communities.

Belonging can be stronger than evidence because religion is tied to family, culture, and memory.

Religious boundaries can preserve identity, but they can also divide people into "us" and "them."

The idea of being chosen can give courage, but it can also create spiritual pride and exclusion.

Leaving religion may carry a heavy social and emotional cost.

Sacred loyalty can sometimes produce moral blindness.

Community, meaning, and compassion can exist without belief in God.

Religion's social usefulness does not prove its supernatural claims.

Transition to the Next Chapter

Belonging made religion powerful among ordinary people. But religion became even more powerful when belief, ritual, and sacred stories were organized into authority.

Someone had to interpret the gods.

Someone had to perform the rituals.

Someone had to define purity.

Someone had to speak for the ancestors.

Someone had to say what God wanted.

From this need came priests, temples, sacred offices, religious law, and spiritual power.

The next chapter examines how religion became authority, and how claims about God became instruments of social and political control.

Sources

Durkheim, Émile. The Elementary Forms of Religious Life. 1912.

Haidt, Jonathan. The Righteous Mind: Why Good People Are Divided by Politics and Religion. Pantheon, 2012.

Norenzayan, Ara. Big Gods: How Religion Transformed Cooperation and Conflict. Princeton University Press, 2013.

Smith, Christian. Religion: What It Is, How It Works, and Why It Matters. Princeton University Press, 2017.

Whitehouse, Harvey. The Ritual Animal: Imitation and Cohesion in the Evolution of Social Complexity. Oxford University Press, 2021.

Chapter 9: Priests, Temples, Power, and Sacred Authority

When Someone Claimed to Speak for God

Once human beings believed in gods, spirits, ancestors, invisible forces, sacred laws, and hidden worlds, a new question appeared:

Who understands these powers?

Who knows what the gods want?

Who can interpret dreams?

Who can perform the ritual correctly?

Who can remove impurity?

Who can forgive sin?

Who can bless a marriage?

Who can bury the dead safely?

Who can explain disaster?

Who can speak for God?

From this question came one of the most powerful figures in human history: the religious specialist.

He may have been called shaman, priest, prophet, healer, oracle, monk, guru, elder, holy man, divine messenger, or spiritual guide. The names differed across cultures, but the social function was often similar. This person claimed special knowledge of the invisible world.

That claim changed everything.

The ordinary person feared disease, death, storm, drought, guilt, impurity, and divine anger. The religious specialist claimed to know how to respond. He knew the words. He knew the rituals. He knew the offerings. He knew the sacred stories. He knew which rule had been broken. He knew what the ancestors wanted. He knew which god had been offended. He knew how order could be restored.

This gave him authority.

Not because he discovered a medicine, built a bridge, predicted an eclipse through astronomy, or proved a theory through evidence. His authority came from a different source: sacred access. He stood between the frightened human being and the invisible power.

That position was extraordinary.

A person who speaks for the gods can influence the living. A person who controls rituals can control fear. A person who defines sin can shape conscience. A person who promises forgiveness can direct obedience. A person who claims to know the afterlife can influence how people live before death.

This is how religion moved from belief to institution.

God became more than an idea.

God became authority.

The Shaman at the Edge of Fear

Before there were temples, there may have been shamans, healers, dreamers, visionaries, and ritual experts.

Imagine a small community facing illness, hunger, and danger. Most people are afraid. They do not know what causes disease. They do not know why the hunt failed. They do not know why a child has convulsions, why a woman cannot conceive, why a man hears voices, why the dead appear in dreams, or why a storm destroyed the village.

Then one person claims to know.

He enters trance. He chants. He shakes. He speaks in strange rhythms. He says he has traveled to the spirit world. He says an ancestor is angry. He says a demon must be driven away. He says the animal spirit must be honored. He says the sick child needs a ritual. He says he has seen what ordinary people cannot see.

To the community, this may feel convincing. The shaman appears different. He seems brave enough to enter the invisible world. He seems to carry knowledge others lack. His performance may be dramatic. His words may be emotionally powerful. His rituals may calm the group.

Some shamans may have been sincere. Some may have had unusual psychological experiences. Some may have been skilled observers of people. Some may have known herbs, wounds, animals, weather, and the emotional needs of their community. Some may have helped people. Some may also have misled them. Human history rarely fits simple categories.

The important point is this: religious authority may have begun where fear met performance.

The shaman did not need to prove the spirit world scientifically. He needed to persuade the community emotionally. If the sick recovered, his power seemed confirmed. If the sick died, the explanation could be adjusted: the spirit was too strong, the ritual was too late, the family failed, or the gods had another plan.

This pattern would appear again and again in religious history.

When claims cannot be tested clearly, authority becomes difficult to challenge.

Temples Made Belief Permanent

A shaman may travel.

A priesthood stays.

A ritual may be temporary.

A temple is permanent.

The rise of temples marked a major change in religion. Belief became architecture. Sacred power was placed in stone, wood, gold, altar, statue, shrine, gate, tower, and inner chamber. The invisible became organized around a visible center.

A temple tells people that the sacred is not only in the mind. It has a location. It has a house. It has rules of entry. It has guardians. It has ceremonies. It has wealth. It has authority.

The temple became the place where the human and divine worlds were said to meet.

People brought offerings. Priests performed rituals. Sacred objects were protected. Festivals were organized. Kings sought legitimacy. Families came for blessing. The sick came for healing. The guilty came for purification. The grieving came for comfort.

A temple can be beautiful. It can preserve art, music, memory, community, and solemnity. It can give people a sense of awe. It can remind them that life is larger than private appetite.

But a temple also concentrates power.

Someone controls the altar.

Someone controls the sacred space.

Someone decides who may enter and who must remain outside.

Someone receives offerings.

Someone interprets the rules.

Someone manages the wealth.

Someone claims divine approval.

This is how religion becomes institutional. It is no longer only a story told by grandparents or a prayer whispered by a frightened person. It becomes land, property, hierarchy, ritual schedule, written law, and social influence.

Once religion gains buildings, wealth, and offices, it becomes harder to question. A private belief may change quietly. An institution defends itself. It has interests. It has status. It has followers. It has rules. It has memory. It has power.

Temples made belief visible.

They also made belief durable.

The Power to Explain Suffering

Knowledge gives power.

Medical knowledge gives power over disease. Legal knowledge gives power in court. Scientific knowledge gives power over nature. Technical knowledge gives power over machines. Religious knowledge, when accepted by a community, gives power over fear, guilt, death, and meaning.

The priest did not merely perform rituals.

He interpreted reality.

A plague comes. The priest says why.

A child dies. The priest says what it means.

A king wins a battle. The priest says God favored him.

A woman is infertile. The priest says what ritual is needed.

A person feels guilt. The priest says what sin was committed.

A family suffers misfortune. The priest says what impurity must be removed.

A doubter asks questions. The priest says doubt is dangerous.

This power to interpret is enormous.

Human beings do not live by events alone. They live by the meaning given to events. The same illness can be understood as infection, punishment, bad luck, witchcraft, karma, test, or divine will. The interpretation changes the response.

If illness is infection, people seek treatment.

If illness is punishment, people seek forgiveness.

If illness is possession, people seek exorcism.

If illness is impurity, people seek purification.

If illness is witchcraft, they may accuse a neighbor.

The interpretation can heal or harm.

Religious authority becomes dangerous when interpretation is protected from evidence. If the priest is wrong, who can correct him? If the prophet fails, who can question him? If the sacred text appears cruel, who can challenge it? If suffering is explained as divine will, who can demand a better answer?

This is one of the great problems of sacred authority.

It often places interpretation above investigation.

Science asks the interpreter to prove the claim.

Religion often asks the believer to trust the interpreter.

That difference has shaped history.

Fear, Guilt, and the Obedient Soul

Religious authority often works through three powerful human emotions: fear, guilt, and hope.

Fear asks: What if I disobey?

Guilt asks: What have I done wrong?

Hope asks: How can I be forgiven or saved?

The religious specialist can stand at the center of all three.

He can say what God forbids. He can say what sin means. He can say how punishment may come. He can say how forgiveness may be obtained. He can say who is pure and who is impure. He can say who belongs and who is excluded. He can say what must be confessed, paid, sacrificed, repeated, avoided, or believed.

This gives religious authority an intimate reach.

Political power can control public behavior. Religious power can enter private thought. It can shape desire, shame, sexuality, doubt, memory, and identity. It can make a person feel watched even when alone. It can make a natural feeling become sin. It can make honest doubt feel like betrayal.

This does not mean all religious authority is cruel. Many religious leaders have offered comfort, education, charity, courage, and moral guidance. Many have helped the poor, cared for the sick, and defended justice. Human institutions are complex.

But we must still examine the structure of power.

A system that claims to speak for God can become difficult to resist. To disagree with a human leader is one thing. To disagree with God is another. When a priest, prophet, scripture, or institution is treated as God's voice, human authority becomes sacred authority.

That is when obedience can become dangerous.

A person may obey not because the command is wise, kind, or just, but because it is said to come from God.

History gives many warnings about this.

Kings Needed Gods, and Gods Needed Kings

Political power and religious power often found each other useful.

Kings needed legitimacy. Gods gave it.

Priests needed protection and wealth. Kings provided it.

A ruler who claims divine approval becomes more than a ruler. He becomes chosen, anointed, sacred, or necessary to cosmic order. Rebellion against him may become rebellion against heaven. His laws may be described as divine will. His victories may be celebrated as proof of God's favor. His enemies may be condemned as enemies of the sacred order.

This relationship between throne and altar appears in many societies.

The ruler supports temples.

The priest blesses the ruler.

The ruler protects the priesthood.

The priesthood teaches obedience.

The state enforces religious law.

Religion sanctifies political power.

This arrangement can create stability, but it can also protect injustice. If power is sacred, criticism becomes sin. If hierarchy is divine, equality becomes rebellion. If a ruler is chosen by God, suffering under that ruler may be interpreted as duty.

Human beings have often used God to strengthen human control.

This does not mean every religious person supports tyranny. Many religious movements have challenged oppression. But historically, religion has often been used to make power appear holy.

That matters for the argument of this book.

If God were simply an independent reality discovered by human beings, we might expect religious authority to stand above human politics. Yet again and again, gods appear entangled with kings, empires, land, war, law, taxes, temples, caste, class, and gender.

The divine often carries the fingerprints of power.

Wealth at the Altar

Religion does not live on belief alone.

It often requires resources: land, buildings, offerings, salaries, animals, grain, gold, books, clothing, instruments, candles, oil, incense, food, schools, monasteries, shrines, and festivals.

Where there are offerings, there is economy.

Where there is sacred land, there is property.

Where there are temples, there is administration.

Where there are pilgrims, there is commerce.

Where there are priests, there is support.

Sacred institutions can become wealthy. This wealth may be used for beauty, charity, education, art, music, architecture, feeding the poor, caring for the sick, and preserving culture. But it can also create privilege, corruption, competition, and dependence.

People may give because they love God.

People may give because they fear punishment.

People may give because they seek blessing.

People may give because they feel guilty.

People may give because everyone else gives.

People may give because religious authority demands it.

Once wealth flows toward sacred institutions, questions become necessary.

Who controls it?

Who benefits?

Who is excluded?

Who is told that poverty is holy while religious leaders live comfortably?

Who is promised blessing in exchange for giving?

Who is made afraid not to give?

This is not an attack on generosity. Giving can be noble. Charity can reduce suffering. Communities need support. But when giving is tied to supernatural promises or fears, it must be examined carefully.

Religion can turn economic exchange into sacred duty.

That gives it power over both the wallet and the conscience.

The Sacred Control of Bodies

Religious authority often reaches the body.

What may be eaten.

What may be touched.

Who may marry.

Who may have sex.

When a woman is pure or impure.

What clothing is modest.

What hair may be shown.

What food may enter the mouth.

What words may be spoken.

What rituals must follow birth, menstruation, sex, sickness, or death.

Control over the body becomes control over daily life.

This is especially important because the body is where human beings feel desire, shame, hunger, illness, pleasure, pain, sexuality, aging, and death. If religion controls the body, it controls some of the most intimate parts of life.

Many purity systems likely developed from a mixture of hygiene, social boundary, gender control, fear of disease, fear of sexuality, and symbolic order. Some rules may have had practical value in certain environments. Others may have reinforced hierarchy and exclusion.

The danger begins when a social rule becomes divine command.

Then women may be called impure because of normal biology.

Lower classes may be called spiritually inferior.

Outsiders may be called polluted.

Sexual desire may be treated as shame.

The sick may be blamed.

The body may become a source of guilt instead of understanding.

Modern science gives a different view. Menstruation is biology, not impurity. Disease is pathology, not moral stain. Sexuality is part of human nature, though it requires ethics and responsibility. Food safety is evidence-based, not spiritually magical. Public health is better than taboo.

When knowledge grows, many old purity systems lose their authority.

But they often survive because they are not only about health. They are about identity, obedience, and control.

When Doubt Became Sin

In ordinary life, doubt can be healthy.

A good doctor doubts until the diagnosis is clear. A good scientist doubts until evidence is strong. A good judge doubts before deciding guilt. A good historian doubts before accepting a claim. Doubt protects us from error.

But many religious systems treat doubt as danger.

Doubt may be called sin.

Questioning may be called pride.

Investigation may be called rebellion.

Disagreement may be called corruption.

Changing belief may be called betrayal.

This is understandable from the point of view of religious authority. If people ask too many questions, authority weakens. If sacred claims must face evidence, some may fail. If people compare religions, inherited certainty becomes less stable. If people question the priest, prophet, or scripture, power becomes less secure.

So doubt must be controlled.

The believer may be told to trust. To obey. To surrender. To stop asking. To accept mystery. To fear punishment. To avoid dangerous books. To stay away from doubters. To protect faith like a fragile flame.

This may preserve belief, but it does not prove truth.

A true idea should not need fear to protect it. A sound argument should not require censorship. A reliable claim should survive examination. If a belief collapses when questioned, perhaps the problem is not the question.

Science grows through doubt.

Religion often survives by limiting it.

This is one of the most important differences between scientific authority and sacred authority. Science can be wrong, but its method allows correction. Sacred authority often resists correction because correction may be treated as disobedience.

That is why the freedom to question is essential.

Without it, human beings remain trapped inside inherited answers.

Why People Wanted Sacred Authority

It would be too simple to say that priests and religious leaders gained power only by deception.

The truth is more complicated.

People often wanted religious authority. They wanted someone to guide them. They wanted someone to explain suffering. They wanted someone to bless marriage, comfort grief, bury the dead, interpret guilt, and provide certainty in moments of fear.

A frightened person may welcome authority.

A grieving person may need words.

A guilty person may seek forgiveness.

A confused person may want rules.

A dying person may ask for assurance.

Religious specialists served these needs. Many did so with sincerity. They gave structure to life's hardest moments. They stood beside people at birth, marriage, illness, and death. They preserved stories, rituals, music, and memory. They created continuity across generations.

This human service is real.

But service and truth are not the same.

A person may offer comfort and still teach falsehood. A leader may be sincere and still mistaken. A ritual may help people grieve and still not open a path to another world. A priest may reduce anxiety and still have no special access to God.

This distinction allows us to be fair.

We can acknowledge the human role of religious authority without accepting its supernatural claims.

We can respect the comfort some leaders have given without surrendering the right to question their power.

We can recognize religion's social function while still asking whether God exists.

That balance is important for this book and for the whole series.

From Sacred Power to Human Responsibility

The great shift from religion to science is also a shift in responsibility.

If disease is caused by demons, we need exorcism.

If disease is caused by germs, we need sanitation, vaccines, antibiotics, clean water, nutrition, and medical care.

If drought is punishment, we need repentance.

If drought is climate and weather, we need irrigation, storage, agriculture, planning, and science.

If disaster is divine anger, we need sacrifice.

If disaster is natural process, we need engineering, warning systems, public safety, and compassion.

Sacred authority often explains suffering by looking upward.

Scientific and humanistic thinking asks what can be understood and changed here.

This does not remove every mystery. It does not remove grief. It does not make life easy. But it moves us from submission to responsibility.

That is one of the great moral gains of natural explanation.

When we stop seeing every misfortune as divine message, we can begin to ask better questions.

Who is suffering?

What caused it?

What evidence do we have?

What can be prevented?

What can be treated?

What can be repaired?

What must be changed?

Religion often asks people to obey the sacred order. Science and ethics ask people to understand reality and reduce suffering.

This is why sacred authority must be examined. If a claim about God prevents knowledge, justice, medicine, compassion, or human freedom, then the claim must not be protected merely because it is old.

The age of a belief does not make it true.

The sacredness of an authority does not make it right.

Chapter Summary

As belief in invisible powers grew, human societies created religious specialists: shamans, priests, prophets, healers, ritual experts, and sacred interpreters. These figures claimed special access to gods, spirits, ancestors, rituals, or divine law. This gave them authority over fear, guilt, sickness, death, purity, morality, and social order.

Religious authority often served real human needs. It comforted the grieving, organized rituals, preserved memory, and gave people guidance during life's most frightening moments. But it also created power. Those who claimed to speak for God could influence rulers, control bodies, define sin, collect wealth, limit doubt, and place human rules beyond question.

This chapter has shown how religion moved from belief to institution. God became more than an explanation.

God became authority.

And once authority became sacred, questioning it became dangerous.

Key Points

Religious specialists gained power by claiming special knowledge of the invisible world.

Temples made belief visible, permanent, organized, and wealthy.

Priests and prophets interpreted suffering, disease, disaster, guilt, and morality.

Sacred authority often works through fear, guilt, hope, and obedience.

Religion and political power have often strengthened each other.

Religious institutions can provide comfort and charity, but they can also protect privilege and control.

Purity rules often extend religious authority into the body and daily life.

Doubt is often treated as dangerous because questioning weakens sacred authority.

A leader may be sincere and still mistaken.

Human beings need comfort and guidance, but need does not prove divine truth.

Transition to the Next Chapter

Religious authority gave belief structure, but stories gave it life.

A law can command, but a story can capture the heart. Myths told people where the world came from, why suffering exists, why death came, why a people was chosen, why a ritual matters, and what destiny awaits the faithful.

Human beings are storytelling animals. We remember stories, repeat stories, fight for stories, and live inside stories.

The next chapter examines myths, sacred imagination, and the power of religious stories to make human meaning feel divine.

Sources

Bellah, Robert N. Religion in Human Evolution: From the Paleolithic to the Axial Age. Harvard University Press, 2011.

Durkheim, Émile. The Elementary Forms of Religious Life. 1912.

Eliade, Mircea. The Sacred and the Profane: The Nature of Religion. Harcourt, 1959.

Norenzayan, Ara. Big Gods: How Religion Transformed Cooperation and Conflict. Princeton University Press, 2013.

Weber, Max. The Sociology of Religion. Beacon Press, 1963.

Chapter 10: Myths, Stories, and Sacred Imagination

The Story That Reached the Heart First

Human beings do not live by facts alone.

We live by stories.

A fact may inform the mind, but a story enters memory. A fact may explain one event, but a story can explain a whole life. A fact may be correct, but a story can be loved, repeated, sung, painted, carved into stone, taught to children, and carried across generations.

This is one reason religion became so powerful.

Religion did not spread only through fear, law, priesthood, or ritual. It spread through stories: creation stories, flood stories, stories of ancestors, stories of heroes, stories of prophets, stories of miracles, stories of suffering, stories of punishment, stories of salvation, stories of heaven, stories of hell, and stories of the end of the world.

A story can make the invisible feel near.

A story can make the dead feel present.

A story can make a tribe feel chosen.

A story can make suffering feel meaningful.

A story can make obedience feel noble.

A story can make death feel less final.

A story can turn fear into purpose.

Long before most human beings could read or write, they could remember stories. Around fires, in caves, in fields, in temples, in homes, beside rivers, and under night skies, stories carried human memory. They told people who they were, where they came from, why they suffered, whom they should obey, what they should fear, and what might happen after death.

This chapter is about the power of sacred stories.

It is not enough to ask whether a religious story is comforting. Many are. It is not enough to ask whether a religious story is beautiful. Many are. It is not enough to ask whether a religious story shaped civilization. Many did.

The deeper question is this:

Does the power of a story prove that the story is true?

The answer is no.

A story can be beautiful and false. A story can be meaningful and human-made. A story can preserve wisdom and still contain error. A story can help a community survive and still fail as evidence of God.

Religious myths are among humanity's most powerful creations. To understand religion, we must understand why human beings needed these stories, why they survived, and why they came to feel sacred.

Mystery Needed a Plot

Mystery is difficult to carry.

A child asks, "Where did the world come from?"

A grieving person asks, "Why did death take my loved one?"

A suffering person asks, "Why do bad things happen?"

A community asks, "Why are we here?"

A tribe asks, "Why are we different from others?"

A ruler asks, "Why should people obey?"

A dying person asks, "What comes next?"

These questions are large. They are emotionally heavy. They do not sit easily in silence. Human beings often answer such questions by building stories around them.

A myth gives mystery a plot.

It says: in the beginning, this happened.

It says: the gods acted.

It says: humans disobeyed.

It says: death entered the world.

It says: the flood came.

It says: the chosen ancestor received a promise.

It says: the prophet heard a voice.

It says: the savior came.

It says: final judgment will come.

The story gives direction to what otherwise feels like chaos. It gives a beginning, a conflict, a meaning, and sometimes an ending. It tells people that life is not random. It belongs to a larger pattern.

This is psychologically powerful.

A person can endure pain more easily if the pain has a place in a story. A community can endure hardship if hardship is part of divine testing. A people can survive exile if exile is part of sacred history. A death can feel less cruel if death is not the end of the plot.

Human beings do not simply ask what happened. They ask what it means.

Myths answer that second question.

Science answers differently. Science asks what caused an event, how it happened, whether it can be measured, whether it can be tested, and whether the explanation can be corrected. Myth asks how the event fits into meaning.

Both may arise from human need, but they are not the same kind of explanation.

A myth may give emotional meaning without giving factual truth.

That distinction is essential.

Creation Stories Gave Humanity a Beginning

Almost every culture has asked where the world came from.

Before cosmology, geology, evolutionary biology, physics, and astronomy, human beings looked at the world and wondered: how did all this begin?

The sky.

The Earth.

The waters.

The animals.

The human body.

The first man.

The first woman.

The first death.

The first law.

The first fire.

The first wrong.

The first people.

Creation stories answered these questions. Some described gods shaping the world from chaos. Some described the world emerging from water, darkness, cosmic egg, divine speech, sacrifice, battle, or birth. Some imagined the Earth as mother. Some imagined the sky as father. Some imagined humans formed from clay, dust, blood, breath, wood, or divine intention.

These stories differ greatly, but they share a human need: the need for origin.

We want to know where things came from. We are not satisfied with the world simply being here. We ask for a beginning because beginnings help us understand identity. A child asks, "Where did I come from?" A nation asks, "Where did we begin?" A religion asks, "How did creation start?"

Creation stories made the world feel less accidental.

They also placed human beings inside cosmic meaning. The world was not merely matter. It was made, ordered, blessed, cursed, tested, or assigned purpose. Human beings were not merely animals. They were created, chosen, commanded, fallen, or destined.

This made creation stories emotionally attractive.

But the existence of many different creation stories creates a serious problem. They cannot all be literally true. If one story says the world emerged from cosmic waters, another says it came from divine speech, another says it was born from a cosmic egg, another says it came from a battle among gods, and another says it was formed from the body of a slain being, we must ask: why does humanity produce so many different beginnings?

The simplest answer is that creation stories reflect human imagination, culture, environment, and need.

They tell us how people understood themselves.

They do not necessarily tell us how the universe began.

Science later offered a different kind of origin story. It did not begin with a sacred ancestor or divine command. It began with observation, measurement, mathematics, cosmic expansion, background radiation, stars, chemistry, geology, biology, and deep time.

Science did not give humanity a comforting childhood story. It gave us a harder, stranger, wider picture.

The universe is not a village with a sky roof. The Earth is not the center of creation. Human beings are not placed at the beginning. We arrive very late in a long cosmic history.

That scientific story is less flattering.

But it is far better supported by evidence.

Suffering Needed a Reason

One of the hardest things for human beings to accept is innocent suffering.

A child becomes sick.

A good person loses everything.

A village is destroyed.

A mother dies in childbirth.

A family is betrayed.

A kind man is murdered.

A cruel ruler prospers.

Without a story, suffering can feel unbearable. It can seem meaningless, random, and insulting to the human sense of justice.

Religious stories often give suffering a reason.

They say suffering entered the world because humans disobeyed.

They say suffering is punishment.

They say suffering is a test.

They say suffering purifies the soul.

They say suffering will be rewarded.

They say suffering is part of a divine plan.

They say the present life is not the whole story.

These answers can comfort people. They can help the suffering endure. They can give language to pain. They can turn despair into patience.

But they can also harm.

If suffering is always interpreted as deserved, victims may be blamed. If illness is punishment, the sick may feel shame. If poverty is destiny, injustice may be ignored. If abuse is explained as spiritual testing, the abused may be told to remain silent. If everything is part of God's plan, people may become less urgent in fighting preventable suffering.

A story can comfort, but it can also excuse.

This is why the meaning given to suffering must be examined carefully.

Science does not answer suffering by giving it cosmic purpose. Medicine asks what disease is present. Public health asks what can be prevented. Psychology asks how trauma affects the mind. Law asks who caused harm. Ethics asks how suffering can be reduced.

Science may not make suffering meaningful in the religious sense, but it can make suffering more understandable and often more preventable.

This is a moral improvement.

It is kinder to treat infection than to call it punishment. It is better to build earthquake-resistant structures than to blame victims. It is better to confront injustice than to tell the poor their suffering is spiritually useful.

Religious myths gave suffering a plot.

Science and ethics give suffering a response.

That may be less poetic, but it is often more humane.

Heroes and Prophets Gave Hope a Face

Religious stories are often built around extraordinary figures.

A prophet hears the divine voice.

A hero defeats evil.

A savior suffers for others.

A sage discovers truth.

A king is chosen.

A martyr dies faithfully.

A holy person performs miracles.

A teacher gives law.

These figures give a human face to sacred meaning. They show people how to live, what to value, what to fear, and what to hope for. A community may see its highest ideals in them.

Heroes and prophets are powerful because human beings learn through example. A rule may command, but a story embodies the rule. A doctrine may be abstract, but a person in a story makes it memorable.

Do not merely be brave. Be like the hero.

Do not merely obey. Follow the prophet.

Do not merely endure suffering. Remember the martyr.

Do not merely hope. Trust the savior.

Sacred figures become emotional anchors. Children learn their names. Families repeat their stories. Artists paint them. Temples and churches honor them. Songs praise them. Their words become law. Their lives become patterns for millions.

This is not surprising. Human beings are social learners. We imitate, admire, remember, and idealize. A powerful figure can shape moral imagination across centuries.

But here again we must distinguish meaning from evidence.

A story about a holy figure may inspire courage, compassion, sacrifice, or reform. That does not prove every supernatural claim attached to the figure. A person may have existed historically, yet legends around that person may grow over time. A wise teacher may become miracle worker in later memory. A local prophet may become cosmic savior. A political reformer may become divine messenger. A martyr may become proof of heavenly reward.

Human memory does not preserve stories like a camera. It shapes them.

Communities remember what they need. They add meaning. They smooth contradictions. They highlight courage. They enlarge events. They protect sacred reputation. Over time, a human life can become a sacred legend.

This does not mean every religious story is deliberately false. Often, legend grows because people love, admire, and need the figure. Devotion can reshape memory without conscious fraud.

That is one reason sacred stories must be studied historically, not only devotionally.

The historian asks: what can we reasonably know?

The believer often asks: what does this story mean for faith?

Those are different questions.

Sacred Memory Became Identity

A group without memory is fragile.

Sacred stories preserve memory. They tell a people who they are. They may describe migration, suffering, liberation, covenant, exile, battle, promise, revelation, or survival. They turn ordinary history into sacred history.

This gives the group strength.

A people may survive hardship because it believes its suffering has meaning. A minority may preserve identity because it carries sacred stories. A community may teach children courage by telling them what ancestors endured. A nation may connect itself to divine destiny. A family may preserve customs because the customs are tied to holy memory.

Sacred memory can be beautiful. It can help people endure. It can resist oppression. It can give dignity to the weak and hope to the defeated.

But sacred memory can also become dangerous when it becomes immune to correction.

If a group believes its history is divinely approved, it may ignore the suffering of others. If it sees itself as always innocent, it may deny its own cruelty. If it sees itself as chosen, it may treat outsiders as lesser. If ancient victory is treated as God's will, modern violence may be justified by old stories.

History asks for evidence.

Myth asks for loyalty.

When myth becomes identity, questioning the myth can feel like attacking the people. The question is no longer, "What happened?" It becomes, "Are you with us or against us?"

That is why sacred stories are so powerful and so difficult.

They hold the group together, but they can also trap the group inside one version of memory.

A mature society must learn to honor memory without turning memory into unquestionable myth.

The past matters.

But the past must still face truth.

How a Story Becomes Sacred

How does an ordinary story become sacred?

It is repeated.

It is tied to ritual.

It is taught to children.

It is linked to ancestors.

It is protected by authority.

It is recited at festivals.

It is written in holy language.

It is surrounded by music, art, architecture, and moral emotion.

It is said to come from God.

Over time, the story becomes more than a story. It becomes a sacred container of identity. People do not merely hear it. They kneel before it. They sing it. They fear misquoting it. They kiss the book that contains it. They build buildings around it. They argue over its meaning. They may fight and die for it.

This transformation is one of the most important processes in religion.

A story becomes sacred when a community gives it ultimate authority.

Once sacred, the story becomes difficult to question. If there are contradictions, they must be explained away. If the story contains cruelty, it may be defended as divine justice. If modern knowledge challenges it, knowledge may be rejected. If morality develops beyond it, morality may be accused of rebellion.

This creates a problem for truth.

A story protected from criticism may survive, but survival is not proof. Many false beliefs survive because they are emotionally protected, socially repeated, and institutionally defended.

Science works differently.

A scientific claim does not become true because it is old, loved, repeated, or sacred. It must face evidence. It must be open to correction. It must survive testing.

This is why science has been so disruptive to religious myths. Science does not ask whether a story is beloved. It asks whether the story matches reality.

That question can be painful.

But without it, human beings remain prisoners of inherited stories.

Stories Travel Faster Than Careful Facts

Stories are biologically and socially powerful.

They are easier to remember than lists. They carry emotion. They create characters. They offer conflict. They produce suspense. They give moral lessons. They connect past, present, and future. They help people teach children.

A fact may say: disease can spread through contaminated water.

A story says: a village angered the gods, and plague came.

The second is easier to remember in a prescientific culture. It has drama, guilt, punishment, and warning. It creates moral urgency.

A fact may say: drought can result from climate variation.

A story says: the rain god withheld mercy because the people failed in ritual.

Again, the story is emotionally stronger.

This is one reason myths survived for generations. They were not stored in textbooks. They lived in memory, song, ritual, and fear. They were carried by the emotional brain. A good myth could travel farther than a careful explanation.

Even today, stories often defeat facts in public life. Rumors spread faster than corrections. Conspiracy theories attract more attention than statistical analysis. A personal anecdote may feel more convincing than a controlled study. A dramatic testimony may move people more than a careful report.

Religion understood this long before modern psychology did.

It placed its teachings inside stories.

Stories of creation.

Stories of sin.

Stories of miracles.

Stories of punishment.

Stories of rescue.

Stories of judgment.

Stories of eternal life.

This does not make religion foolish. It makes religion humanly intelligent. It uses the way the mind works. It speaks in images, not only arguments. It reaches the heart before the mind has time to object.

For a belief system to survive, it must be memorable.

Myth is memorable.

That is one reason myth became sacred.

Sacred Imagination Created Beauty and Danger

Sacred imagination has created great beauty.

It has inspired architecture, music, painting, poetry, sculpture, festivals, moral language, funeral rituals, and acts of compassion. It has helped human beings face grief. It has given communities meaning. It has given the poor a sense of dignity. It has preserved human questions about life, death, suffering, justice, and hope.

We should not deny this.

Some atheist writing becomes weak because it treats religion only as stupidity or poison. That is too simple. Religion is also imagination, longing, memory, art, fear, tenderness, and moral struggle. It contains cruelty, but also beauty. It contains superstition, but also poetry. It contains control, but also comfort.

A serious criticism of religion must be honest enough to see both.

But beauty does not prove truth.

A cathedral can be beautiful without proving that its theology is true.

A hymn can be moving without proving heaven.

A sacred poem can be profound without proving revelation.

A myth can contain wisdom without being history.

A ritual can be meaningful without being supernatural.

This distinction allows us to appreciate human creativity without surrendering intellectual honesty.

The danger begins when sacred imagination is mistaken for divine fact. Then stories can become weapons. Myths can justify exclusion. Poetry can become dogma. Allegory can become law. Ancient imagination can control modern life.

A beautiful story may inspire us.

But it should not be allowed to imprison us.

From Sacred Story to Human Meaning

If religious stories are human creations, does life lose meaning?

No.

It means human beings are meaning-makers.

We create meaning through memory, love, work, art, family, justice, service, curiosity, courage, and compassion. We tell stories because we are conscious beings living in time. We remember the past. We imagine the future. We suffer. We love. We lose. We hope.

The problem is not that humans create meaning. That is part of our dignity.

The problem begins when created meaning is mistaken for external supernatural fact.

A funeral story may help a family grieve. That does not prove an afterlife.

A creation myth may help a people understand identity. That does not prove how the universe began.

A miracle story may inspire faith. That does not prove the event occurred as told.

A story of divine law may shape morality. That does not prove the law came from God.

We do not need to destroy all meaning in order to reject supernatural claims. We can read myths as human documents: windows into fear, hope, morality, imagination, and culture. We can learn from them without kneeling before them.

This is one of the goals of this book.

To understand religion as human creation is not to make humanity smaller. It may make humanity more understandable. We see our ancestors trying to answer the impossible with the tools they had. We see them frightened under the sky, grieving at graves, dreaming of the dead, telling stories to children, and trying to make life bearable.

God may have been the greatest story human beings ever told.

But a great story is still not the same as truth.

Chapter Summary

Human beings are storytelling animals. Before science provided evidence-based explanations, myths gave mystery a plot. They explained creation, suffering, death, morality, identity, and destiny. Sacred stories helped communities preserve memory, teach children, strengthen belonging, and survive hardship.

Religious myths became powerful because they were emotionally rich, memorable, repeated, ritualized, and protected by authority. They gave human beings meaning in the face of fear and uncertainty. But meaning is not the same as evidence. A story can be beautiful, useful, and morally moving without being literally true.

This chapter has shown that sacred imagination helped create and preserve belief in God. Myths did not merely describe religion. They made religion live inside memory, culture, and identity. But when myths are mistaken for unquestionable truth, they can also resist correction, justify power, and prevent honest inquiry.

God may be humanity's most powerful story.

But truth requires more than a powerful story.

Key Points

Human beings remember and transmit stories more easily than abstract facts.

Myths give mystery a plot and make uncertainty emotionally manageable.

Creation stories reflect the human need for origin and identity.

Religious stories often give suffering a reason, but this can comfort or harm.

Heroes, prophets, saviors, and martyrs give sacred meaning a human face.

Sacred stories preserve group memory and identity.

A story becomes sacred when it is repeated, ritualized, protected, and treated as divine.

Stories may survive better than facts because they carry emotion.

Sacred imagination can create beauty, but beauty does not prove truth.

Human beings can find meaning without mistaking myth for evidence.

Transition to the Next Chapter

Sacred stories were powerful, but they were not the same everywhere.

Different lands produced different gods. Different fears produced different rituals. Different climates, rivers, mountains, wars, foods, rulers, and histories shaped different religious imaginations.

If religion came from one clear divine source, we might expect one clear message. Instead, human history shows many gods, many myths, many scriptures, many rituals, and many contradictions.

The next chapter examines why different cultures created different gods.

Sources

Armstrong, Karen. A Short History of Myth. Canongate, 2005.

Campbell, Joseph. The Hero with a Thousand Faces. Princeton University Press, 1949.

Dundes, Alan, editor. Sacred Narrative: Readings in the Theory of Myth. University of California Press, 1984.

Eliade, Mircea. Myth and Reality. Harper & Row, 1963.

Leeming, David A. Creation Myths of the World: An Encyclopedia. ABC-CLIO, 2010.

Chapter 11: Why Different Cultures Created Different Gods

A World of Many Heavens

If one clear, universal God had spoken plainly to all humanity, we might expect human religion to show one clear message.

One God.

One truth.

One moral law.

One description of the afterlife.

One path to salvation.

One explanation of creation.

One reliable revelation for all people in all places.

But that is not what human history shows.

Instead, we find many gods, many spirits, many scriptures, many rituals, many heavens, many hells, many creation stories, many sacred laws, many prophets, many saviors, many priesthoods, many taboos, and many contradictions.

Some people worshiped the Sun.

Some worshiped the moon.

Some worshiped rivers.

Some worshiped mountains.

Some worshiped ancestors.

Some worshiped many gods.

Some worshiped one God.

Some imagined God as father.

Some imagined the divine as mother.

Some believed in reincarnation.

Some believed in resurrection.

Some believed in heaven and hell.

Some believed in liberation from rebirth.

Some believed God demanded sacrifice.

Some believed God demanded purity.

Some believed God demanded obedience to law.

Some believed the divine was beyond all images and names.

This variety is not a small detail. It is one of the strongest clues that religions are shaped by human life.

Religious beliefs did not appear in one uniform form. They grew from soil, climate, geography, language, fear, food, war, family, politics, memory, and imagination. The gods of desert people often carried the concerns of desert life. The gods of farming people often cared about rain, fertility, harvest, and soil. The gods of seafaring people often ruled waves, storms, and dangerous journeys. The gods of warrior cultures often blessed battle, courage, conquest, and tribal loyalty.

The map of belief followed the map of life.

This does not, by itself, settle every philosophical question about God. But it strongly suggests that human beings did not simply discover one clear divine reality. They created religious worlds that reflected their own earthly worlds.

To understand gods, we must study the people who imagined them.

Geography Wrote Theology

Geography shapes imagination.

A desert does not teach the same lessons as a rainforest. A river valley does not create the same fears as a mountain village. An island people do not face the same dangers as a landlocked tribe. A farming civilization does not live with the same urgency as a nomadic herding group.

The land enters religion.

In dry places, rain becomes sacred. Water becomes blessing. Wells become life. A god who controls rain is not a minor god. He is the difference between survival and death.

In river civilizations, the river may become divine because it floods, feeds, fertilizes, and destroys. The Nile, the Ganges, the Tigris, the Euphrates, the Indus, and other great rivers did not merely carry water. They carried civilization. It is not surprising that rivers became sacred.

In mountains, height itself can feel holy. Mountains rise above ordinary life. Clouds gather around them. Storms form there. Their peaks seem closer to the sky. Many cultures imagined mountains as places where gods lived, spoke, descended, or revealed law.

In forests, spirits may seem hidden among trees, animals, shadows, and sounds. In the sea, gods may become unpredictable, violent, generous, and dangerous, like the water itself. In volcanic regions, fire beneath the earth may become a sign of underworld power or divine anger.

The environment teaches what a people fears and what a people needs.

A god of rain matters most where rain is uncertain.

A god of the sea matters most where the sea rules life.

A god of fertility matters most where crops and children determine survival.

A god of war matters most where enemies threaten the group.

A god of law matters most where social order becomes complex.

This pattern is deeply human. People look at the forces that control their lives and imagine sacred powers behind those forces.

The gods become local because human life is local.

Hunger Created Fertility Gods

Before supermarkets, refrigeration, tractors, antibiotics, irrigation systems, and modern transport, survival was fragile.

Food was never guaranteed.

A failed harvest could kill. A diseased herd could ruin a family. A flood could destroy stored grain. Insects could devour crops. Drought could empty the land. A hard winter could turn hunger into death.

In such a world, food was not just food.

It was life.

This helps explain why so many religions developed rituals around planting, harvest, animals, fertility, first fruits, fasting, feasting, and offerings. The divine world was asked to protect the food supply because the food supply was the future of the group.

Agricultural societies often imagined gods and goddesses of grain, rain, soil, fertility, seasons, and harvest. Herding societies cared deeply about animal fertility, pasture, milk, and protection from disease. Hunting societies often developed rituals of respect toward animals, as if the animal world had spiritual agency.

The body and the land were connected.

Human fertility and agricultural fertility often became religiously linked. The womb and the field both produced life. Blood, seed, soil, rain, milk, and harvest became powerful symbols. Birth and farming were surrounded by awe because both involved hidden life emerging into the visible world.

Before biology and agriculture were understood scientifically, these processes felt mysterious. How does a seed become grain? How does a child form in the womb? Why does one woman conceive and another not? Why does one field flourish and another fail? Why does one herd multiply and another weaken?

Religion answered through sacred fertility, blessing, purity, ritual, and divine favor.

Again, the pattern is visible: human need created religious meaning.

The gods did not float above human survival. They were tied to bread, milk, rain, soil, animals, sex, birth, and hunger.

A starving people does not first ask for abstract theology.

It asks for food.

So it creates gods who can feed it.

War Put a Flag in Heaven

Human beings have always lived with conflict.

Groups competed for land, water, animals, crops, trade routes, honor, safety, and power. Before modern states, courts, and international agreements, group survival often depended on loyalty and defense. A weak group could be raided, enslaved, displaced, or destroyed.

In such a world, religion easily became tribal.

Our god protects us.

Our god gave us this land.

Our god hates our enemies.

Our god demands courage.

Our god blesses our warriors.

Our god explains our victory.

Our god explains our defeat.

This kind of religion strengthens group unity. It gives courage to soldiers. It makes sacrifice meaningful. It tells the community that war is not merely political or economic, but sacred.

The danger is obvious.

If a war is blessed by God, then ordinary human enemies become divine enemies. Killing can become obedience. Conquest can become destiny. Cruelty can become purification. Victory can become proof of righteousness. Defeat can become punishment for sin.

This does not mean religion always causes war. Human beings fight for land, food, power, fear, pride, revenge, and survival. But religion can intensify conflict by giving it cosmic meaning.

The god of war often looks like the tribe that needs him.

He carries its flag.

He speaks its language.

He remembers its enemies.

He blesses its dead.

This is not what we would expect from one clear universal truth. It is exactly what we would expect from human beings creating sacred stories around their own struggle to survive.

Kings Gave Heaven a Throne

Political organization also shaped God.

Small tribes may imagine spirits, ancestors, and local powers. Larger kingdoms and empires often imagine more organized heavens. As human society develops hierarchy, divine society often develops hierarchy too.

Earthly kings sit on thrones. Heavenly gods sit on thrones.

Earthly kings command armies. Heavenly gods command angels, spirits, or divine messengers.

Earthly kings issue laws. Heavenly gods issue commandments.

Earthly kings demand loyalty. Heavenly gods demand worship.

Earthly kings punish rebellion. Heavenly gods punish sin.

This is not accidental. Human beings often imagine the unseen world through the social world they know.

As societies became more centralized, religion often became more centralized. A powerful empire might favor a powerful high god. A king might claim divine support. Priests might protect royal authority. Royal power and religious power could strengthen each other.

This was not only theology.

It was politics.

A ruler blessed by God becomes harder to challenge. A law said to come from heaven becomes harder to revise. A hierarchy declared sacred becomes harder to question. A priesthood tied to the ruler can teach obedience as virtue.

This helps explain why God often appears as king, judge, lawgiver, and ruler.

These images did not arise in a vacuum. They grew from human systems of authority.

If a people lives under empire, heaven may look imperial.

If a people lives under monarchy, heaven may look royal.

If a people lives under law, heaven may look legal.

If a people lives under patriarchy, heaven may look male.

The divine image often follows the structure of human power.

That does not prove that no God exists in every possible philosophical sense. But it does make many religious images of God look human-made.

Language Gave God a Local Voice

Religion is carried by language.

A sacred word, prayer, chant, scripture, hymn, or name of God does not exist outside human speech. It enters the mind through sound and meaning. It carries the rhythm of a people. It preserves their metaphors, memories, fears, and hopes.

This is why God often speaks the language of the believer.

A child hears God in the language of home. Sacred words become tied to mother, father, grandmother, festival, prayer, and community. Even when translated, religious language often carries the emotional weight of its original culture.

Language shapes theology.

A word for spirit may also mean breath.

A word for sin may suggest debt, stain, rebellion, impurity, or disobedience.

A word for God may suggest father, lord, ruler, power, sky, ancestor, or being.

A word for salvation may mean rescue, liberation, healing, release, forgiveness, or union.

Different languages do not merely label the same ideas. They shape how people imagine the sacred.

This matters because divine revelation is often claimed to be universal, yet it arrives in particular languages, through particular cultures, at particular times. Sacred texts carry grammar, metaphor, politics, gender assumptions, social customs, and historical concerns of the people who produced and preserved them.

If God's message were universal and clear, why would it be so dependent on ancient languages, disputed translations, cultural context, and human interpretation?

This question becomes stronger when we see how religious groups disagree over the meaning of their own sacred texts. The words may be revered, but interpretation remains human. Scholars, priests, sects, reformers, and believers argue over meaning because language is not self-explaining.

God, in practice, speaks through human tongues.

And human tongues carry human history.

Birthplace Is the Strongest Missionary

Perhaps the most powerful argument in this chapter is also the simplest:

Birthplace predicts belief.

Most people do not choose their religion after carefully studying all religions. They inherit the religion of family and society. A child born in Saudi Arabia is far more likely to become Muslim than Hindu. A child born in India is far more likely to become Hindu than Mormon. A child born in Utah is more likely to encounter Mormon belief. A child born into a secular Scandinavian family may grow up with little religious commitment. A child born into an Orthodox Jewish family is likely to inherit Jewish practice. A child born into a Catholic village, a Buddhist community, a Sikh family, or an evangelical household will usually receive that world before being able to compare it with others.

This should make every honest person pause.

If belief were mainly a matter of independent evidence, geography would not matter so much. But it matters enormously.

The religion that feels obvious to one person may feel strange to another simply because they were born in different homes. The sacred story that moves one heart may leave another unmoved because it was not part of childhood memory. The prayer that feels natural in one language may feel foreign in another.

This does not mean every believer is insincere. Most are sincere. But sincerity often reflects inheritance, not discovery.

A person may say, “I know my religion is true.” But millions of equally sincere people in other traditions say the same thing. They cannot all be right in their specific claims. They can all feel certain, but certainty is not the same as truth.

Birthplace is a powerful teacher.

Family is a powerful teacher.

Community is a powerful teacher.

Repetition is a powerful teacher.

This does not prove atheism by itself, but it weakens religious certainty. If the belief one holds is strongly predicted by birth, then humility is required.

One should ask:

Would I believe this if I had been born elsewhere?

That question has changed many lives.

Contradictions Beneath the Sacred Surface

The diversity of religion would be less troubling if religions differed only in poetry, rituals, and symbols.

But they often contradict each other on major claims.

Is there one God or many?

Is God personal or impersonal?

Does God have form or no form?

Is there heaven and hell?

Is there reincarnation?

Is there one life or many lives?

Is salvation through faith, works, ritual, grace, knowledge, devotion, obedience, liberation, or enlightenment?

Was the world created once, created repeatedly, emanated, shaped, born, spoken into being, or eternal?

Are humans fallen, ignorant, tested, trapped, chosen, sinful, divine, or spiritually evolving?

Which scripture, if any, is final?

Which prophet, if any, speaks for God?

Which rituals are necessary?

Which foods are forbidden?

Which moral laws are eternal?

These are not small differences.

Religions make competing claims about reality. They cannot all be literally true in the way their strongest defenders claim. At most, one could say they are different human attempts to approach mystery. But that is very different from saying one clear God gave one clear truth.

The contradictions matter because they suggest a human source.

When human cultures create stories, we expect variation. When people imagine gods from local needs, we expect many gods. When religious authority grows in different places, we expect different rules. When myths develop over generations, we expect contradictions.

Religious diversity is exactly what we would expect if religion is human-made.

It is not what we would expect if one all-powerful, loving God wanted to reveal one clear saving truth to all humanity.

The believer may answer that God revealed truth, but humans misunderstood it. Yet that answer creates another question: why would an all-knowing God communicate in a way so easily misunderstood, fragmented, and culturally confined?

The confusion of religion is itself evidence that religion is human.

Gods Changed When People Changed

Gods are often presented as eternal.

But religious ideas change.

Gods merge.

Gods disappear.

Gods gain new qualities.

Gods lose old qualities.

Local gods become universal gods.

War gods become moral gods.

Storm gods become creator gods.

Fertility goddesses become symbolic figures.

Tribal gods become cosmic rulers.

Old rituals are reinterpreted.

Old laws are softened, ignored, or defended selectively.

Religions adapt because human societies adapt.

When people move from hunting to farming, religion changes. When villages become cities, religion changes. When kingdoms become empires, religion changes.

When literacy spreads, religion changes. When philosophy develops, religion changes. When science challenges old explanations, religion changes. When moral consciousness shifts, religion changes.

This is why many modern believers do not practice religion exactly as their ancestors did. They may reinterpret ancient texts. They may reject old punishments. They may soften ideas of hell. They may emphasize love over fear. They may read myths symbolically rather than literally. They may accept science while preserving faith.

This flexibility shows human agency.

Religions survive by changing. But if a belief changes repeatedly under pressure from culture, ethics, politics, and science, it becomes harder to claim that it represents one unchanging divine truth.

The history of religion is not a straight line from God to humanity.

It is a human conversation across time.

It includes fear, power, poetry, reform, conflict, compromise, and imagination.

God often changes when humans change because God is often a reflection of human life.

The Outsider's Honest Question

Every believer should sometimes view his or her religion as an outsider.

This is difficult, but necessary.

A Christian should ask: how does Christianity look to someone raised outside it?

A Muslim should ask: how does Islam look to someone raised outside it?

A Hindu should ask: how does Hinduism look to someone raised outside it?

A Jew, Buddhist, Sikh, Jain, Mormon, tribal believer, and atheist should all ask similar questions.

From inside a tradition, the belief feels natural. From outside, it looks like one human tradition among many.

This outside view is powerful because it removes the privilege we give to our own inheritance. It asks us to apply the same standards to our own religion that we apply to others.

If I reject another religion's miracle stories because they are based on testimony, should I accept my own miracle stories on similar testimony?

If I see another scripture as culturally shaped, should I examine whether mine is also culturally shaped?

If I think another group's rituals are human inventions, why do I assume mine came from God?

If I see that another person's certainty came from childhood teaching, should I ask whether mine did too?

This is not an attack.

It is intellectual fairness.

We must not use one standard for other people's beliefs and another standard for our own.

The outsider's question is dangerous to inherited certainty. But it is necessary for honest thinking.

A belief that cannot survive being viewed from the outside may not be as strong as it feels from the inside.

What Religious Diversity Really Shows

Religious diversity is often celebrated as beautiful, and in many ways it is. It shows human creativity, poetry, music, ritual, symbolism, longing, and moral imagination. It shows how deeply human beings have struggled with death, suffering, nature, guilt, hope, and meaning.

But religious diversity also creates a serious challenge to divine truth claims.

If each culture creates gods that reflect its own life, then gods begin to look like cultural products.

If birthplace predicts belief, then certainty begins to look inherited.

If religions contradict each other, then at least many must be wrong in their literal claims.

If gods change when societies change, then theology begins to look historical.

If sacred texts require endless human interpretation, then revelation begins to look less clear than believers claim.

None of this requires hatred of religion. In fact, it allows a more humane understanding of religion. We can see religions as human attempts to answer the same deep problems: fear, death, nature, morality, suffering, social order, and hope.

But we should not confuse human attempts with divine answers.

Different cultures created different gods because different cultures lived different lives.

That is the central argument of this chapter.

Religion is not best understood as one clear truth descending from heaven. It is better understood as many human responses rising from Earth.

Chapter Summary

Different cultures created different gods because they lived in different environments, faced different dangers, spoke different languages, organized power differently, and inherited different histories. Desert people, river civilizations, mountain communities, farming societies, seafaring groups, warrior cultures, and empires all shaped religion according to their needs and fears.

Religious diversity is one of the strongest clues that gods are human cultural creations. If one clear God had revealed one clear truth to all humanity, we might expect religion to be more consistent. Instead, we see many gods, many rituals, many scriptures, many afterlife beliefs, many moral codes, and many contradictions.

Birthplace strongly predicts belief. Most people inherit religion before they examine it. This does not make believers dishonest, but it does weaken the claim that religious certainty is based mainly on evidence. The variety and contradiction of religious belief suggest that religion comes from human culture more than divine revelation.

Different gods did not fall from one heaven.

They rose from different human worlds.

Key Points

Geography, climate, food, danger, and survival needs shaped religious imagination.

Agricultural societies often developed fertility, rain, harvest, and soil-related religious ideas.

War and tribal conflict helped create protective and warrior gods.

Political systems shaped gods as kings, rulers, judges, and lawgivers.

Language carries religious meaning and makes revelation dependent on human culture.

Birthplace is one of the strongest predictors of religious belief.

Religions contradict one another on major claims about God, salvation, afterlife, scripture, and morality.

Gods and religious ideas change when societies change.

The outsider's view helps people examine inherited belief more fairly.

Religious diversity suggests human creation more strongly than one clear divine revelation.

Transition to the Next Chapter

Different cultures created different gods, but individuals usually do not begin by comparing cultures.

They begin as children.

A child does not first ask for evidence. A child trusts. A child listens. A child repeats prayers, watches rituals, absorbs fear, learns sacred stories, and accepts the world given by family.

This is how religion enters the mind before reason is fully formed.

The next chapter examines childhood, family, and inherited belief.

Sources

Armstrong, Karen. A History of God: The 4,000-Year Quest of Judaism, Christianity and Islam. Knopf, 1993.

Geertz, Clifford. The Interpretation of Cultures. Basic Books, 1973.

Hume, David. The Natural History of Religion. 1757.

Pew Research Center. The Global Religious Landscape. Pew Research Center, 2012.

Smart, Ninian. The World's Religions. Cambridge University Press, 1998.

Chapter 12: Childhood, Family, and Inherited Belief

The Child Who Believes Before He Can Question

No child is born reading scripture.

No child enters the world asking whether the universe needs a creator, whether miracles are reliable, whether the soul can survive the death of the brain, or whether one religion is more likely to be true than another.

A child enters the world helpless.

The child needs food, warmth, protection, touch, language, and love. The child does not survive by skepticism. The child survives by trust.

If the mother says, "Do not touch the fire," the child should listen.

If the father says, "Stay away from the road," the child should obey.

If the family says, "This food is safe," the child accepts.

If the elders say, "This place is dangerous," the child believes.

Trust is not a weakness in childhood. It is a survival tool.

This is one reason religion enters the mind so easily. It comes through the same voices that teach the child language, safety, kindness, manners, family history, and daily life. The same mother who gives food may teach prayer. The same father who protects from danger may speak of God. The same grandmother who tells stories of the family may also tell stories of gods, prophets, spirits, heaven, hell, karma, sin, blessing, and punishment.

To the child, these teachings do not arrive as uncertain theories.

They arrive as reality.

The child does not say, "This is one possible religious tradition among many."

The child says, "This is how the world is."

That is the beginning of inherited belief.

Religion becomes powerful not only because it is taught, but because it is taught early, tenderly, repeatedly, and by the people the child loves most.

Before the child can examine religion, religion has already entered memory.

Before the child can compare beliefs, one belief has already become home.

The Family Becomes the First Temple

For many people, the first temple is not a building.

It is the family.

It is the mother's prayer. It is the father's instruction. It is the grandparent's story. It is the lamp lit in the evening, the song before sleep, the holiday meal, the sacred object on the wall, the religious book on the shelf, the festival clothes, the rules about food, the words spoken at birth, and the rituals performed at death.

Religion often enters the child not as doctrine, but as atmosphere.

The child watches adults bow, kneel, chant, fold hands, light candles, touch feet, cover heads, remove shoes, fast, feast, wear symbols, avoid certain foods, and speak certain words with reverence. Long before the child understands theology, the child understands seriousness.

Some things are ordinary.

Some things are sacred.

Some places are casual.

Some places require silence.

Some words can be spoken casually.

Some words must be spoken with care.

This is how religion becomes embodied. The body learns before the mind analyzes. The child learns posture, rhythm, gesture, and tone. The child learns when to be quiet, when to repeat, when to fear, when to celebrate, when to feel guilty, and when to feel proud.

Later, the adult may think of religion as belief. But for the child, religion begins as life.

This is why inherited religion can be so difficult to question. To question religion may feel like questioning the family itself. The belief is tied to the smell of food, the sound of songs, the touch of a grandparent's hand, the memory of festivals, the grief of funerals, the blessing before an exam, the prayer during illness, and the stories told with love.

A belief tied to love becomes emotionally protected.

That protection may feel beautiful, but it can also make honest examination difficult.

Sacred Stories Arrive Before Critical Thinking

Children understand stories before they understand evidence.

A child can remember a story long before the child can evaluate whether the story is historically true, scientifically possible, or morally consistent. This is one reason sacred stories are so powerful in childhood.

A child hears that God made the world.

A child hears that God watches everything.

A child hears that prayers are heard.

A child hears that good people go to heaven.

A child hears that bad people are punished.

A child hears that one scripture is holy.

A child hears that one prophet, teacher, savior, or tradition is special.

These ideas enter early. They become part of the child's emotional map. The child may not understand metaphor, myth, history, translation, cultural context, or competing religious claims. The child simply trusts the storyteller.

This is not because children are foolish. It is because childhood is designed for learning. Children absorb language without demanding proof. They absorb customs without studying anthropology. They absorb family loyalty without reading sociology. They absorb religion in the same way.

The problem is that religious claims are not ordinary claims.

If a child is taught the name of a fruit, the claim can be checked. If a child is taught that fire burns, the claim can be verified. If a child is taught that a road is dangerous, real consequences exist.

But if a child is taught that God is watching, that hell exists, that the family's religion is the only truth, or that doubt is sinful, the child has no easy way to test the claim. The claim enters the mind through trust, not evidence.

By the time the child becomes capable of critical thought, the belief may already feel sacred.

This is how religion gains a long head start over reason.

It arrives first.

Fear Enters Early and Stays Deep

Some religious teaching enters through love.

Some enters through fear.

Fear is especially powerful in childhood.

A child who is told that God sees everything may feel watched even when alone. A child who is told about hell may fear punishment more terrible than anything in ordinary life. A child who is told that doubt is sin may become afraid of his own thoughts. A child who is told that certain desires are impure may feel shame before understanding his own body.

These fears can remain for decades.

An adult may stop believing intellectually, yet still feel fear emotionally. The mind may say, "This is not true," while the body still feels anxiety. This happens because early emotional learning can be deep. A warning heard in childhood may remain long after the reason for the warning has been questioned.

Religion often understands this power.

Teach the child early.

Repeat the story.

Connect belief with reward.

Connect doubt with danger.

Connect obedience with love.

Connect disobedience with guilt.

Connect the family's faith with goodness.

Connect outsiders with error.

This creates a powerful emotional structure. The child may grow into an adult who feels that questioning religion is not merely intellectual. It feels dangerous. It feels immoral. It feels like stepping outside protection.

This is one reason many adults avoid religious questioning. They are not only afraid of being wrong. They are afraid of what they were taught would happen if they question.

Fear can protect belief from examination.

A belief protected by fear may last a lifetime, but fear does not make it true.

Love Makes Belief Feel Safe

Fear is powerful, but love may be even more powerful.

Many people remain attached to religion because religion is tied to the people they love. The mother who prayed was loving. The father who taught scripture was sincere. The grandmother who told sacred stories was kind. The community that gathered for worship also brought food during illness, support during grief, and joy during festivals.

Because the people were good, the belief feels good.

This is emotionally understandable.

If a loving person teaches a belief, the belief may feel trustworthy. If a whole family lives inside a religion, the religion feels safe. If religious rituals are tied to birthdays, weddings, funerals, meals, songs, and moral teaching, then questioning religion may feel like rejecting love itself.

But we must be careful.

Good people can believe mistaken things.

A loving parent can pass on an inherited error.

A kind grandparent can teach a superstition.

A sincere community can preserve a false belief.

Love proves love.

It does not prove doctrine.

This distinction is important for anyone questioning inherited religion. One can love the people who gave the belief without accepting the belief itself. One can honor parents without agreeing with every religious claim they taught. One can cherish childhood memories without treating childhood teaching as final truth.

This is not easy. But it is necessary for intellectual freedom.

If we accept every belief because it came from people we love, then we surrender our ability to examine reality.

Love should not require us to stop thinking.

Repetition Turns Teaching Into Truth

A belief repeated often enough begins to feel obvious.

This is true not only in religion. It is true in politics, advertising, family stories, national identity, social customs, and personal habits. Repetition gives an idea familiarity. Familiarity can feel like truth.

Religion uses repetition with extraordinary skill.

Daily prayers.

Weekly worship.

Annual festivals.

Life-cycle rituals.

Sacred songs.

Repeated scriptures.

Common phrases.

Family teachings.

Public symbols.

Repeated warnings.

Repeated promises.

The child hears the same ideas again and again. God is real. Our religion is true. This book is holy. These rituals matter. These people are chosen. These actions are sinful. This prayer protects. This festival is sacred. This doubt is dangerous.

Over time, the belief becomes part of the mind's furniture.

It is no longer experienced as something learned. It feels like something known.

This is why many adults say, "I have always known God exists." Often, what they mean is, "I was taught this so early and so often that I cannot remember not believing it."

Repetition also works through the body. Kneeling, bowing, singing, fasting, feasting, touching sacred objects, wearing religious symbols, and attending rituals all train the body to feel that the belief is real.

The body remembers what the mind has not examined.

This does not make religion false by itself. But it explains why belief can feel certain even when evidence is weak. Certainty may come from repetition, not truth.

The question is not, “How long have I believed this?”

The question is, “What evidence supports it?”

Every Child Grows Inside a Room

Every child grows up inside a room.

Not only a physical room, but a mental room. The room contains language, food, customs, stories, fears, jokes, songs, rules, heroes, villains, and assumptions about the world. For many children, religion is part of the walls of that room.

The child does not see the walls at first.

The room is simply reality.

A child raised in one religious room may learn that one God is the only truth.

A child raised in another room may learn that many gods are natural.

One child may learn that images are sacred.

Another may learn that images are forbidden.

One may learn that eating certain foods is sinful.

Another may learn that the same foods are ordinary.

One may learn that salvation comes through faith.

Another may learn that liberation comes through knowledge, devotion, discipline, karma, grace, or enlightenment.

To each child, the room feels normal.

This is the power of culture. It does not merely teach answers. It teaches what questions are allowed. It teaches which doubts feel dangerous. It teaches which outsiders seem strange. It teaches what emotions should arise around God, death, sex, purity, authority, and truth.

Growing intellectually means noticing the room.

It means asking: what did my culture teach me before I could choose? What did my family assume? What did my community protect? What did I absorb before I could compare?

This is not an act of hatred.

It is an act of maturity.

A person cannot think freely until he can see the room in which his thinking began.

Doubt Can Feel Like Betrayal

For many people, religious doubt hurts because it feels disloyal.

The doubter may feel he is betraying his parents. She may feel she is breaking her grandmother's heart. He may fear dishonoring ancestors. She may fear losing community.

He may worry that his family will think he has become arrogant, immoral, corrupted, or lost.

This emotional pressure is real.

Religion often becomes tied to family honor. A child may be praised for faith and warned against doubt. The family may treat religious obedience as a sign of goodness. Belief becomes not only a claim about reality, but a measure of loyalty.

This makes honest questioning difficult.

A person may ask privately: what if I do not believe this anymore?

Then another thought appears: what will this do to my family?

That second question can silence the first.

Many people continue religious practices long after belief has weakened because leaving would cause pain. Some remain silent. Some pretend. Some perform rituals for family peace. Some teach children beliefs they themselves doubt because they fear conflict. Some live divided lives: outward obedience, inward uncertainty.

This shows again that religion is not only an idea.

It is a social bond.

To question it can feel like tearing a fabric.

But truth cannot be decided by fear of disappointing others. A belief does not become true because parents love it. A doctrine does not become true because ancestors preserved it. A ritual does not become true because leaving it hurts.

Doubt may feel like disloyalty, but honest questioning is not betrayal.

It is the beginning of intellectual adulthood.

Inheritance Must Face Examination

Every person inherits many things: language, food habits, family stories, cultural identity, political assumptions, moral instincts, and religious beliefs.

Inheritance is not automatically bad.

We need inheritance. Without it, each generation would begin from nothing. We learn from those before us. We receive wisdom, language, affection, memory, and practical knowledge.

But inheritance must eventually face examination.

Some inherited beliefs are good.

Some are harmless.

Some are beautiful.

Some are mistaken.

Some are harmful.

A mature person must learn the difference.

This is especially important with religion because religious beliefs make large claims. They speak about the origin of the universe, the purpose of life, the fate of the dead, the nature of morality, the meaning of suffering, the rules of sexuality, the value of doubt, and the destiny of the soul.

These are too important to accept only because they were inherited.

A person may begin with the religion of childhood. That is understandable. But an adult has the responsibility to ask whether the belief is true.

Does it have evidence?

Would I believe it if I were born elsewhere?

Does it explain reality better than natural explanations?

Does it require fear to protect it?

Does it survive comparison with other religions?

Does it make moral sense?

Does it help reduce suffering, or does it create unnecessary guilt and division?

These questions do not insult childhood.

They honor truth.

The goal is not to erase memory. The goal is to examine belief.

We may keep gratitude for the people who raised us, while still refusing to let childhood decide reality for us.

The Courage to Think Beyond Childhood

Leaving inherited certainty can be frightening.

When a person begins to question childhood religion, the world may feel unstable. Old answers weaken. Familiar rituals lose authority. Prayers may feel like words spoken into silence. Sacred stories may begin to look human. Fear may rise. Guilt may return. The person may wonder, "If this is not true, what else have I believed without evidence?"

This can be painful.

But it can also be liberating.

To think beyond childhood is not to reject everything loved. It is to become honest about what is known, what is hoped, what is feared, and what is unsupported. It is to stop confusing early teaching with final truth.

The child had to trust.

The adult must examine.

This is a natural movement in human life. We do not continue believing everything we were told as children. We learn that some stories were simplified. Some warnings were exaggerated. Some customs were local. Some fears were unnecessary. Some authorities were wrong. Some traditions were beautiful but not factual.

Religious belief deserves the same examination.

A person may examine religion and still remain religious. Another may become agnostic. Another may become atheist. The important point is not that everyone must reach the same conclusion immediately. The important point is that inherited belief should not be protected from honest questioning.

A belief worthy of respect should not fear examination.

The courage to think beyond childhood is one of the first steps toward intellectual freedom.

Chapter Summary

Religious belief often begins in childhood, before critical thinking is fully developed. Children survive by trust. They accept what parents, grandparents, and communities teach them. Religion enters through love, fear, repetition, ritual, stories, festivals, songs, and family identity. By the time the adult mind becomes ready to question, the belief may already feel like truth.

This chapter has shown that most people do not discover religion independently. They inherit it. Birthplace, family, language, and community strongly shape belief. This does not make believers dishonest, but it does weaken the claim that religious certainty is based mainly on evidence.

Inherited belief can be emotionally powerful because it is tied to love, memory, and belonging. But love does not prove doctrine. Repetition does not prove truth. Childhood teaching does not decide reality. Mature thinking requires us to examine what we received before we could question.

The child must trust to survive.

The adult must question to become free.

Key Points

Children trust before they question, and this makes early religious teaching powerful.

The family often functions as the first temple.

Sacred stories enter the mind before the child can evaluate evidence.

Fear of hell, punishment, impurity, or divine surveillance can shape belief deeply.

Love makes inherited belief feel safe, but loving teachers can still pass on mistaken ideas.

Repetition can make belief feel obvious even when evidence is weak.

Every child grows up inside a cultural and religious room.

Doubt often feels like disloyalty because religion is tied to family and identity.

Inherited belief should eventually face adult examination.

The child must trust, but the adult must think.

Transition to the Next Chapter

Religion enters early through family, fear, love, and repetition. But it does not remain only in childhood. It continues to feel real throughout life because it is supported by emotion, memory, ritual, music, community, coincidence, and the deep human need for meaning.

A person may say, "I know God is real because I feel it."

That feeling deserves attention.

But feeling real and being real are not the same.

The next chapter examines why belief feels so real, and why emotional certainty is not the same as evidence.

Sources

Barrett, Justin L. Born Believers: The Science of Children's Religious Belief. Free Press, 2012.

Harris, Paul L. Trusting What You're Told: How Children Learn from Others. Harvard University Press, 2012.

Kelemen, Deborah. "Are Children Intuitive Theists? Reasoning About Purpose and Design in Nature." Psychological Science, 2004.

Richert, Rebekah A., and Paul L. Harris. "The Ghost in My Body: Children's Developing Concept of the Soul." Journal of Cognition and Culture, 2006.

Whitehouse, Harvey. The Ritual Animal: Imitation and Cohesion in the Evolution of Social Complexity. Oxford University Press, 2021.

Chapter 13: Why Belief Feels So Real

The Feeling That Becomes Proof

Many believers do not begin with an argument.

They begin with a feeling.

They say, "I know God is real because I feel His presence."

That feeling may come during prayer, music, grief, danger, silence, worship, nature, illness, recovery, childbirth, death, or sudden relief. It may come in a temple, church, mosque, shrine, hospital room, bedroom, forest, or lonely night. It may arrive as peace, warmth, awe, comfort, strength, forgiveness, or the sense that one is not alone.

To the person who feels it, the experience can be overwhelming.

It may feel more convincing than logic.

It may feel more personal than science.

It may feel more certain than evidence.

This is one reason religion is so durable. Religious belief is not only taught in words. It is felt in the body. The heart becomes calm. Tears come. Fear softens. The music rises. A crowd sings. A candle burns. A familiar prayer is repeated. A memory returns. A suffering person feels held by something larger than himself.

In that moment, God feels real.

But we must ask a careful question.

Does feeling prove reality?

The answer must be no.

A dream can feel real. A memory can feel certain and still be wrong. A fear can feel justified and still be exaggerated. A coincidence can feel meaningful and still be accidental. A crowd can make an emotion stronger without making the belief true.

The human mind does not only observe the world. It creates meaning. It fills silence with voice, grief with presence, fear with protection, and uncertainty with story.

This chapter does not deny that religious feelings are real as experiences. They are real to the person having them. But the question is different: do those feelings prove that God exists outside the mind?

That is a much harder claim.

Emotion Speaks Before Reason Arrives

Emotion is faster than reasoning.

Before we carefully analyze a situation, the body has often already reacted. Fear tightens the chest. Hope lifts the mind. Shame burns. Grief weakens the body. Love softens the face. Anger sharpens the voice. Awe stills the heart.

Human beings do not live as pure reasoning machines.

We feel first.

Then we explain.

This matters deeply in religion.

A person who feels peace after prayer may conclude that God answered. A person who feels guilt after breaking a rule may conclude that God is displeased. A person who feels awe under the stars may conclude that the universe must have a divine maker. A person who feels comfort during grief may conclude that the dead still live. A person who feels fear after doubt may conclude that doubt itself is dangerous.

The feeling comes first.

The interpretation follows.

But the same emotional experience can be interpreted differently in different cultures. A Christian may feel the presence of Jesus. A Hindu may feel the blessing of a deity. A Muslim may feel closeness to Allah. A Buddhist may feel peace without a creator God. A person in an ancestor tradition may feel the presence of the dead. A secular person may feel awe, beauty, gratitude, or love without interpreting it as divine.

The feeling may be similar.

The explanation changes with culture.

This is important. If the feeling itself proved one specific religion, then similar emotional states would not lead different believers to different religious conclusions. Yet they do. This suggests that the feeling is human, while the religious interpretation is learned.

Emotion may show that something matters to us.

It does not automatically show what is true about the universe.

Music Turns Belief Into a Bodily Experience

Music is one of the most powerful tools of religion.

A sacred song can open the heart before the mind has time to examine the words. A chant can slow breathing. A drumbeat can unite bodies. A hymn can bring tears. A repeated prayer can quiet anxiety. A group singing together can make each person feel part of something larger.

Religion understood the power of music long before modern science studied emotion, rhythm, and the brain.

Music moves through the body. It does not ask permission from logic. It can produce awe, sadness, joy, courage, surrender, and belonging. In a religious setting, this emotional power is often interpreted as the presence of God.

The believer may say, “I felt God during worship.”

But perhaps the person felt music, memory, rhythm, group emotion, expectation, and longing.

This does not make the experience fake. It may be deeply moving. It may even change the person’s life. But emotional intensity does not prove supernatural origin.

A patriotic song can bring tears.

A love song can awaken memory.

A funeral song can break the heart.

A lullaby can calm a child.

A national anthem can make a crowd feel united.

A religious hymn works through the same human body and brain. It may be attached to sacred words, but its emotional power is not mysterious. Human beings are musical, rhythmic, social creatures. We respond to melody, repetition, harmony, and shared voice.

This helps explain why religious belief feels real in the body. Belief is not only thought. It is sung, breathed, repeated, heard, and felt.

When an idea is carried by music, it enters deeper than argument.

That makes it powerful.

It does not make it proven.

Memory Gives God a Familiar Voice

Memory is not a perfect recording.

It is a living reconstruction.

Every time we remember, we rebuild. We do not simply open a stored file. We reassemble pieces of emotion, image, meaning, and story. Memory can be warm, distorted, selective, comforting, painful, or incomplete.

Religion often lives inside memory.

A prayer heard in childhood.

A grandmother's voice.

A father's blessing.

A festival light.

A funeral chant.

A sacred song.

A smell of incense.

A temple bell.

A church organ.

A holy book touched with reverence.

These memories make belief feel familiar. And familiarity can feel like truth.

A person may not believe because of philosophical argument. He may believe because God is connected with the deepest emotional memories of childhood. God is in the room where the family prayed. God is in the words spoken over the dead. God is in the food prepared during holy days. God is in the face of a mother who taught faith with love.

To question God can then feel like questioning memory itself.

But memory is not evidence of supernatural reality. It is evidence of human attachment. It shows that belief has been woven into the emotional life of the person.

This is why religious belief can remain strong even after arguments weaken. The belief is protected by memory. It is not only in the head. It is in childhood, grief, family, and identity.

A person may say, "This belief feels true because it has always been with me."

But something can feel true because it is old inside us.

Age inside the mind is not proof outside the mind.

Coincidence Wears the Clothing of Meaning

A person prays for help, and help comes.

A family asks for a sign, and something unusual happens.

A patient is told the prognosis is poor, then improves.

A man misses a train, then learns there was an accident.

A mother feels sudden fear for her son, then later hears he was in danger.

A dream seems to match a later event.

Such moments can feel astonishing. They seem too meaningful to be random. They seem arranged, directed, personal. They can become the private evidence by which a person believes in God.

But coincidence is common in a world filled with countless events.

Every day, billions of people think, dream, worry, pray, travel, fall sick, recover, make choices, avoid danger, meet strangers, and remember unusual moments. With so many events occurring, some will line up in surprising ways. The human mind then notices the meaningful matches and forgets the many non-matches.

This is how coincidence becomes testimony.

The prayer followed by recovery is remembered. The prayers followed by death are explained away.

The dream that seemed predictive is retold. The thousands of dreams that predicted nothing are forgotten.

The near accident avoided by delay is called protection. The fatal accident after prayer is called mystery.

The mind is not dishonest. It is selective. It naturally gives weight to emotionally striking events.

Religion often grows through these remembered coincidences. Families preserve them. Communities repeat them. A coincidence becomes a sign. A sign becomes testimony. Testimony becomes proof in the mind of the believer.

Science does not deny that coincidences happen. It asks whether they occur more often than chance, whether they can be tested, whether they can predict future events, and whether alternative explanations have been ruled out.

A memorable coincidence may change a life.

But by itself, it does not prove God.

The Mind Remembers the Hits and Forgets the Misses

Human beings tend to notice what supports what they already believe.

This is called confirmation bias.

A believer may notice every event that seems to confirm God's presence and overlook events that challenge it. An answered prayer is remembered as evidence. An unanswered prayer is explained as God's will. A recovery is called miracle. A death is called mystery. A feeling of peace is called presence. A feeling of absence is called a test.

In this way, belief becomes difficult to disprove.

Whatever happens can be fitted into the belief.

This is not unique to religion. People do this in politics, sports, medicine, business, family disputes, and personal habits. We protect beliefs that matter to us. We look for supporting evidence. We avoid painful contradictions. We defend what feels like part of our identity.

But in religion, confirmation bias can be especially strong because the belief is tied to family, morality, death, hope, and belonging.

A person who wants God to be real may see God everywhere.

In beauty.

In survival.

In coincidence.

In recovery.

In timing.

In emotion.

In scripture.

In silence.

In suffering.

In everything.

The problem is that a belief that explains everything may explain very little. If every outcome confirms the belief, then no outcome can test it.

Science tries to protect us from this. It asks us to look not only for confirming evidence, but also for disconfirming evidence. It asks what would count against the claim. It asks whether the belief can fail a test.

This is why science can correct itself.

A religious belief protected by confirmation bias often cannot.

The Crowd That Makes Certainty Stronger

Belief becomes stronger in a crowd.

A person alone may doubt. But when surrounded by hundreds or thousands of people singing, praying, weeping, clapping, kneeling, chanting, or listening to a powerful speaker, doubt may fade. The emotion of the group enters the individual.

This is not weakness.

It is human social psychology.

We are affected by others. A crowd can make fear stronger, courage stronger, anger stronger, joy stronger, grief stronger, and belief stronger. When many people act as if something is true, it becomes harder to stand apart.

The believer may think, "All these people feel it. All these people know it. Surely this must be real."

But shared emotion is not the same as shared truth.

Crowds have believed many things that were false. Crowds have supported false rumors, unjust wars, cruel traditions, political myths, racial hatred, conspiracy theories, and failed prophecies. A crowd can intensify conviction without improving evidence.

Religious gatherings can be beautiful. They can create comfort, discipline, charity, and emotional release. But they can also produce a feeling of certainty that is socially generated.

This explains why people may return from religious events feeling renewed in faith. The renewal may be real as an emotional state. But it may come from music, rhythm, authority, repetition, social pressure, and shared feeling.

The group can make belief feel larger than the individual.

But truth is not decided by volume.

A thousand voices can sing a falsehood beautifully.

Peace Is Not Proof

One of the most common claims for religion is inner peace.

A person prays and feels calm.

A person surrenders to God and feels relieved.

A person believes in heaven and fears death less.

A person confesses and feels forgiven.

A person trusts divine plan and feels less anxious.

These experiences matter. Peace is valuable. Human beings suffer greatly from fear, guilt, grief, and uncertainty. Anything that brings calm may feel like rescue.

But peace is not proof.

A person may feel peace after accepting a belief that is not true. A person may feel peace from meditation, music, therapy, friendship, exercise, medication, nature, sleep, forgiveness, or simply making a decision. Relief can come when mental conflict ends, even if the conclusion reached is mistaken.

For example, a person who joins a strict group may feel peaceful because uncertainty has been removed. A person who accepts a conspiracy theory may feel peaceful because confusion has been replaced by a clear story. A person who believes a loved one is watching from heaven may feel peaceful because grief has been softened.

The peace is real.

The belief still needs evidence.

This distinction is important because many people treat peace as divine confirmation. They say, "I felt peace, so God must have spoken." But the human nervous system can produce peace through many pathways. Calm does not identify its own cause.

As a physician, I know that symptoms and feelings require interpretation. A patient may feel better for many reasons. The feeling matters, but it does not by itself prove the diagnosis.

In the same way, spiritual peace may be psychologically meaningful without being supernatural evidence.

A belief should not be accepted as true simply because it reduces anxiety.

Suffering Can Make God Feel Necessary

Pain can deepen belief.

A person in comfort may not think much about God. But when illness, grief, failure, poverty, fear, or death comes close, the need for God may become urgent. Suffering pushes the mind toward help. It makes hope feel necessary. It makes prayer feel natural.

This is understandable.

When life is stable, we may tolerate uncertainty. When life collapses, uncertainty becomes unbearable. The suffering person wants meaning, protection, rescue, and assurance. Religion offers all of these.

It says your pain is seen.

It says you are not alone.

It says there is a purpose.

It says justice will come.

It says the dead are safe.

It says God will carry you.

For many people, this belief gives strength. It may help them endure surgery, grief, addiction, loneliness, prison, war, or injustice. We should not mock this. Suffering is heavy, and human beings need support.

But suffering can also make people more vulnerable to unsupported claims.

A desperate person may believe what a calm person would question. A frightened patient may accept miracle promises. A grieving parent may accept messages from the dead. A lonely person may join a group that offers love in exchange for belief. A dying person may accept certainty because uncertainty is too painful.

This does not mean the person is foolish.

It means suffering lowers the tolerance for doubt.

Religion often grows strongest where life is most fragile.

Again, this explains belief.

It does not prove belief.

Pain may make God feel necessary.

But what feels necessary is not always real.

The Brain Can Create Presence

Many people have experienced a sense of presence.

Someone feels there is another person in the room, though no one is visible. A grieving person feels the dead loved one nearby. A person under stress feels watched. A person in prayer feels accompanied. A person in danger feels protected by an unseen power.

Such experiences can be intense.

The sense of presence may feel more than imagination. It may feel external, personal, and undeniable. For the believer, this may become proof of God, angels, spirits, ancestors, or the dead.

But the brain is capable of producing the feeling of presence.

Human beings are social creatures. The brain is built to track other people, infer unseen minds, and detect possible agents. Under grief, stress, isolation, fatigue, fear, ritual, or expectation, the sense of another presence can become strong.

This does not make the experience meaningless. It may reveal grief, need, attachment, fear, or comfort. But it does not automatically prove an external being.

The brain can create a felt presence just as it can create a dream, memory, image, voice, or emotion. The experience may arise inside the mind while being felt as outside the self.

This is one of the most important lessons in understanding religion.

Not everything that feels external comes from outside.

A person may sincerely feel God.

But sincerity does not answer the question of origin.

The experience may come from the brain's deep social machinery, shaped by culture and emotion.

Religious Experience Speaks the Language of Culture

If religious experience came from one clear external God, we might expect such experiences to be more consistent across humanity.

But religious experiences often follow culture.

A Christian may feel the presence of Jesus.

A Catholic may see Mary.

A Hindu may experience Krishna, Shiva, Kali, or another deity.

A Muslim may feel the nearness of Allah without images.

A Buddhist may experience deep compassion, emptiness, or awakening without a creator God.

A person in an ancestor-based culture may feel the dead watching.

A person in a spirit tradition may interpret visions as spirits.

This cultural pattern matters.

The mind has powerful experiences, but culture tells the mind what those experiences mean. A feeling of peace becomes the Holy Spirit in one tradition, divine blessing in another, spiritual energy in another, ancestral presence in another, and nonreligious awe in another.

The experience is real as experience.

The interpretation is learned.

This helps explain why religions are so diverse and why people are often certain in contradictory ways. Each person's experience is filtered through the beliefs already taught by family and culture.

A person raised in one religion rarely has spontaneous experiences confirming another religion. Most religious experience confirms what the person already knows or expects.

This does not prove that all such experiences are false, but it strongly suggests that human culture plays a major role.

God often speaks with the accent of childhood.

Feeling Real Is Not Being Real

This chapter has returned again and again to one central distinction:

Feeling real is not the same as being real.

This may sound simple, but it is difficult to accept because feelings are powerful. We live through them. We trust them. We build identity around them. A strong feeling can seem more convincing than careful evidence.

But human beings can be deeply convinced and deeply mistaken.

People have felt certain about omens.

They have felt certain about witchcraft.

They have felt certain about astrology.

They have felt certain about false cures.

They have felt certain about conspiracy theories.

They have felt certain about political myths.

They have felt certain about dreams, visions, signs, and prophecies.

Certainty is a state of mind, not a guarantee of truth.

This is why science matters. Science does not ask us to stop feeling. It asks us not to treat feeling as final proof. It gives us methods to test what intuition cannot safely decide.

Religious belief often feels real because it is supported by many human forces: emotion, memory, music, ritual, childhood, community, coincidence, fear, grief, hope, and the brain's sense of presence.

These forces are powerful.

They explain why people believe.

They do not prove that the belief is true.

Chapter Summary

Religious belief often feels real because it is supported by deep human emotions and experiences. Prayer can bring peace. Music can create awe. Ritual can organize feeling. Memory can make belief familiar. Coincidence can feel like a sign. Community can strengthen certainty. Suffering can make God feel necessary. The brain can create a sense of presence. Culture then gives religious meaning to these experiences.

This chapter has shown that religious experiences may be sincere and emotionally powerful without proving supernatural reality. The key distinction is between experience and interpretation. A person may truly feel peace, presence, awe, or comfort. But the cause of that feeling may be psychological, social, cultural, or neurological rather than divine.

Belief feels real because it grows from real human needs and real human experiences.

But feeling real is not the same as being true.

Key Points

Many believers treat religious feeling as proof of God.

Emotion often speaks before reason.

Music, rhythm, ritual, and group worship can make belief feel bodily real.

Memory connects religion with childhood, family, grief, and identity.

Coincidence can feel like divine communication.

Confirmation bias helps believers remember confirming events and explain away contradictions.

Crowds can intensify certainty without improving evidence.

Inner peace is valuable, but it is not proof of supernatural truth.

Suffering can make belief feel more necessary.

The brain can create a powerful sense of unseen presence.

Religious experience often follows culture, suggesting that interpretation is learned.

Feeling real is not the same as being real.

Transition to the Next Chapter

We have now seen how fear, death, nature, dreams, ritual, authority, culture, childhood, and emotion all helped make God feel real to human beings.

The final chapter of this book will bring these threads together.

God was humanity's first great explanation. God helped early people explain the sky, disease, death, suffering, morality, and belonging. But the first explanation is not always the best explanation.

The next chapter asks what happens when humanity's first explanation meets evidence, science, and honest doubt.

Sources

Damasio, Antonio. Descartes' Error: Emotion, Reason, and the Human Brain. Putnam, 1994.

Festinger, Leon. A Theory of Cognitive Dissonance. Stanford University Press, 1957.

James, William. The Varieties of Religious Experience. 1902.

Kahneman, Daniel. Thinking, Fast and Slow. Farrar, Straus and Giroux, 2011.

Sacks, Oliver. Hallucinations. Knopf, 2012.

Chapter 14: God as Humanity's First Explanation

The First Answer in a Dark World

Before science, human beings still had to live.

They had to face storms without meteorology. They had to face disease without medicine. They had to face childbirth without anatomy. They had to face drought without climate science. They had to face earthquakes without geology. They had to face dreams without neuroscience. They had to face death without any clear understanding of the brain.

They had questions, but they did not yet have reliable methods.

They could observe. They could remember. They could tell stories. They could repeat rituals. They could trust elders. They could fear omens. They could watch the sky. They could bury the dead. They could pray.

But they could not yet test nature in the scientific sense.

So, when the world frightened them, they created answers from the tools they had: imagination, fear, memory, grief, pattern-seeking, ritual, and story.

Why does the sky roar?

Why did my child die?

Why did the river flood?

Why did the crops fail?

Why do the dead return in dreams?

Why do good people suffer?

Why are we here?

What happens after death?

For these questions, early human beings did not have laboratories, controlled experiments, astronomy, microbiology, physics, evolutionary biology, or brain science.

They had God.

God became the great answer. God could explain the sky, disease, death, morality, disaster, birth, harvest, victory, defeat, suffering, and hope. God was large enough to fill every empty space because human knowledge was still small.

This chapter brings together the central argument of this book.

God was humanity's first great explanation.

But first explanations are often incomplete. Sometimes they are comforting. Sometimes they are useful.

Sometimes they organize life. But being first does not make them true.

Human progress often begins when the first answer is finally questioned.

Mystery Became Someone

One of the most powerful moves in human imagination was turning mystery into person.

The storm became anger.

The disease became punishment.

The dream became visitation.

The dead became ancestors.

The sky became ruler.

The universe became will.

The unknown became someone.

This changed everything.

A blind force cannot be begged. A physical law cannot be pleased. A virus cannot be shamed. An earthquake cannot forgive. A drought cannot be moved by prayer.

But a person can be approached.

A person can be pleaded with.

A person can be thanked.

A person can be feared.

A person can be obeyed.

A person can show mercy.

By turning mystery into person, human beings made the frightening world emotionally negotiable.

If the gods were angry, perhaps they could be calmed.

If ancestors were watching, perhaps they could be honored.

If disease came from spirits, perhaps spirits could be expelled.

If rain came from divine favor, perhaps favor could be won.

If death was a doorway, perhaps the dead were not lost forever.

God made the world feel less indifferent.

That was the emotional power of the idea.

But this also gives us an important clue. God looked very much like the human beings who imagined him. God wanted loyalty, praise, obedience, sacrifice, purity, and fear. God acted like a father, king, judge, warrior, and protector. God carried human emotions: anger, love, jealousy, mercy, disappointment, and pride.

This human shape is not a small detail. It is one of the strongest signs that God was built from human experience.

If human beings created God, we would expect God to look human.

We would expect God to resemble family, tribe, ruler, law, punishment, and protection.

We would expect different cultures to create different gods.

We would expect gods to change as societies changed.

That is exactly what we see.

The Explanation That Could Never Lose

One reason God became so powerful is that God could explain anything.

A child recovered. God healed.

A child died. God had a plan.

Rain came. God was merciful.

Drought continued. God was displeased.

The army won. God gave victory.

The army lost. God punished sin.

A prayer seemed answered. God listened.

A prayer failed. God's ways were mysterious.

A person felt peace. God was present.

A person felt nothing. God was testing faith.

This kind of explanation is emotionally flexible. It protects belief from disappointment. Whatever happens, God can remain in the story.

But there is a problem.

An explanation that can explain every possible outcome may not truly explain anything.

A good explanation should help us understand why one thing happens rather than another. It should clarify causes. It should give us some way to test whether it is right or wrong. It should risk failure. It should meet reality.

The God explanation often avoids that risk.

If success proves God and failure also protects God, then the belief cannot be tested. If recovery proves divine mercy and death proves divine mystery, then the explanation has become protected from evidence.

This is one reason science eventually became so important.

Science did not allow every result to confirm the same idea. It demanded observation, measurement, comparison, correction, and prediction. It asked not only, "Can we tell a story about this?" It asked, "What actually causes this?"

That question slowly changed the world.

The storm became weather.

The plague became infection.

The eclipse became alignment.

The earthquake became geology.

The harvest became agriculture.

The dream became brain activity.

The diversity of life became evolution.

God explained everything when human beings knew little.

Science began to explain better because it was willing to be tested.

What God Gave the Human Heart

If we want to understand why God survived for so long, we must understand how much the idea gave to human beings.

God explained nature.

God comforted grief.

God softened death.

God gave moral rules.

God created group identity.

God supported authority.

God gave ritual structure.

God made suffering meaningful.

God offered hope to the powerless.

God made people feel watched, loved, judged, and protected.

Few ideas in human history have done so many things at once.

This is why religion became so durable. It was not merely a theory about the universe. It was a complete system of emotional survival. It entered birth, childhood, marriage, illness, guilt, law, community, memory, death, and hope.

A purely intellectual argument rarely defeats an idea tied to so many needs. This is why many people continue to believe even when they cannot defend the belief clearly. They are not holding only a proposition. They are holding comfort, family, meaning, identity, and fear.

To understand this is not to surrender to religion.

It is to understand why religion is powerful.

A person who believes because of grief may not be moved by cosmology alone. A person who believes because of childhood fear may not be freed by one argument. A person who believes because religion is tied to family may not question easily.

This is why the first task is understanding.

Once we understand what God gave human beings, we can ask the harder question: does human need prove divine reality?

It does not.

Need explains belief.

It does not prove truth.

Useful Does Not Mean True

Religion was useful.

This must be admitted honestly.

It helped people face death. It united groups. It gave rituals for grief. It created moral pressure. It preserved stories. It gave people courage. It made suffering feel less meaningless. It gave communities a shared identity. It helped many people endure lives filled with pain, poverty, uncertainty, and loss.

But usefulness is not the same as truth.

A belief can help people survive and still be false.

A ritual can reduce fear and still have no supernatural effect.

A myth can unite a tribe and still not be history.

A prayer can calm the mind and still not reach a divine listener.

A belief in heaven can soften grief and still not prove an afterlife.

A belief in divine justice can comfort the oppressed and still not prove that such justice exists beyond death.

This distinction is essential for the entire No God Required series.

Human beings often judge beliefs by how they feel, how old they are, how many people hold them, or how much comfort they bring. Science asks a different question:

What is the evidence?

A belief may be ancient, but age does not make it true.

A belief may be widespread, but numbers do not make it true.

A belief may be comforting, but comfort does not make it true.

A belief may be meaningful, but meaning does not make it true.

Truth requires more than human need.

This is where religion begins to weaken as an explanation. It tells us a great deal about human beings, but very little that can be reliably shown about God.

The Human Fingerprints on God

If God were a clear universal truth revealed plainly to humanity, religion should show more unity.

But history shows diversity.

Different cultures created different gods because they faced different worlds. Desert people feared drought. River civilizations depended on floods. Farming societies cared about fertility. Warrior cultures honored battle. Empires imagined heavenly kings. Patriarchal societies imagined male gods. Local communities honored local ancestors. Sacred stories grew from local memory.

This diversity matters.

God often looks local.

God speaks local languages.

God follows local customs.

God cares about local food rules.

God protects local tribes.

God blesses local rulers.

God carries local fears.

This is what we would expect if religion is human-made. It is not what we would expect from one clear divine message to all humanity.

The believer may answer that one religion is true and all others are mistaken. But that answer creates another problem. Why would a loving and all-powerful God allow truth to depend so heavily on birthplace, family, language, and culture?

A child's religion is often predicted before the child can even speak.

That should make every honest person pause.

Religions grow like cultures grow. They change, split, merge, reform, compete, and adapt. Their gods change with them. Their rituals change. Their moral teachings change. Their interpretations change. Their scriptures are translated, debated, defended, and reinterpreted.

That looks like human history.

It does not look like one unchanging truth delivered clearly from beyond the world.

The First Answer Arrived Early

The first answer has special power because it arrives early.

The child hears it before doubt matures. The family repeats it before comparison begins. The community practices it before the individual can evaluate it. The songs, rituals, fears, festivals, and sacred stories enter memory before reason can fully examine them.

By adulthood, the first answer may feel like reality itself.

This is why religion often feels self-evident to believers. It was not usually accepted after neutral investigation. It was absorbed as part of life. It came with language, family, food, grief, love, and fear.

A person may say, "I have always known God is real."

But perhaps the person has always been told God is real.

This is a difficult distinction, but an important one.

To question childhood religion can feel cruel because it touches the people who gave us love. But we must not confuse love with evidence. Good parents can pass on false beliefs. Sincere communities can preserve errors. Beautiful traditions can contain mistaken explanations.

A mature mind must be able to say: I honor the people who gave me this belief, but I must still ask whether the belief is true.

This is not betrayal.

It is intellectual adulthood.

The First Answer Met the First Real Method

For thousands of years, God survived because there was no better method.

There were observations, stories, rituals, traditions, and authorities, but not yet modern science. There was no reliable way to separate a true explanation from a comforting explanation. There was no controlled experiment. No germ theory. No geology. No evolutionary biology. No neuroscience. No modern astronomy. No statistical method. No peer review. No laboratory culture. No systematic demand that claims be testable and corrected by evidence.

Science changed that.

Science was not simply a collection of facts. It was a method of humility. It admitted that human beings can be wrong. It recognized that intuition can mislead. It understood that stories can comfort and still be false. It created ways to test claims against reality.

That is why science became so dangerous to religious explanation.

Not because science began as hatred of God, but because science asked better questions.

What causes lightning?

What causes disease?

What causes earthquakes?

What causes seasons?

What causes mental illness?

What causes inheritance?

What causes the diversity of life?

What causes the motion of planets?

Whenever these questions received natural answers, the God explanation lost territory.

This is the theme of the next book in the series: the history of science is the history of God becoming unnecessary as an explanation.

But the foundation begins here. Before God could become unnecessary, God first had to become necessary in the human mind. This book has shown why that happened.

God entered the empty spaces of human understanding.

Science began when human beings learned that empty spaces should be investigated, not worshiped.

What We Lose and What We Gain

Some people fear that if God is understood as a human creation, something precious is lost.

In one sense, that fear is understandable.

We may lose the feeling that someone is watching over us. We may lose the promise that death is not final. We may lose the belief that every injustice will be corrected by divine judgment. We may lose the comfort that suffering has a hidden plan. We may lose the certainty that our tribe, scripture, or tradition has cosmic importance.

Those losses can be emotionally difficult.

But we may also gain something.

We gain honesty.

We gain freedom from fear-based belief.

We gain the courage to face death without illusion.

We gain responsibility for reducing suffering here, instead of explaining it away as divine will.

We gain the ability to build morality on compassion rather than command.

We gain the ability to see all human beings as members of one species, not as rival tribes favored or rejected by different gods.

We gain the right to question.

We gain the humility to admit what we do not know.

We gain the wonder of a universe that does not need to be supernatural in order to be magnificent.

Understanding God as a human creation does not make life empty. It makes life more honestly human.

The sky is still beautiful without a sky god.

Love is still meaningful without eternal reward.

Grief is still sacred without heaven.

Morality is still necessary without divine surveillance.

Wonder is still possible without worship.

Meaning is still real when human beings create it through love, truth, service, memory, and compassion.

A Respectful Farewell to the First Explanation

There is no need to hate the first explanation.

We can understand why it arose. We can understand why it comforted. We can understand why it survived. We can understand why people still hold it with deep emotion.

But understanding is not the same as believing.

Humanity once needed God to explain what it did not understand. Today, we no longer need that explanation in the same way. We have better tools. We have better methods. We have better evidence. We have science, history, psychology, medicine, anthropology, neuroscience, and philosophy.

These do not answer every question. They do not remove every sorrow. They do not make human life simple. But they are more honest than inventing invisible answers where evidence is missing.

God was humanity's first great explanation.

But the first explanation was not the final explanation.

Science is not an insult to our ancestors. It is the continuation of their questions. They tried to understand the world with the tools they had. We must do the same with better tools.

The difference is that science no longer stops at the first comforting answer. It asks for evidence. It accepts correction. It allows mystery, but does not worship ignorance.

That is why God, as an explanation, began to lose necessity.

Not because human beings stopped asking deep questions.

But because they found better ways to answer them.

Chapter Summary

God was humanity's first great explanation for fear, death, nature, suffering, morality, dreams, social order, and meaning. Before science, human beings did not have reliable natural explanations for disease, storms, drought, earthquakes, eclipses, fertility, mental experience, or death. God filled these gaps by turning mystery into person, chance into plan, suffering into purpose, and death into passage.

This chapter has brought together the central argument of the book: God was psychologically natural, historically understandable, socially useful, and emotionally powerful. But usefulness is not truth. Comfort is not evidence. Cultural survival is not divine proof. The fact that human beings needed God does not mean God exists.

God was humanity's first explanation.

Science would later show that first explanations can be replaced by better explanations.

Key Points

God became the great early answer to fear, death, mystery, nature, morality, and suffering.

Human beings often turned mystery into person because a personal force could be prayed to, pleased, feared, or obeyed.

The God explanation could explain everything, but often in a way that could not be tested.

Religion served many human needs: comfort, identity, morality, ritual, authority, and hope.

A belief can be useful without being true.

Religious diversity suggests that gods were shaped by local cultures and human needs.

Inherited belief often feels true because it arrives early in childhood.

Science became powerful because it offered a method for testing explanations.

Understanding God as a human creation can lead not to emptiness, but to honesty, freedom, responsibility, and wonder.

Transition to the Conclusion

This chapter completes the main argument of Book 1. We have seen how fear, death, nature, dreams, pattern-seeking, ritual, authority, culture, childhood, emotion, and story helped human beings create God.

But one question remains before we leave this book:

If God was humanity's first explanation, what should we do now?

The conclusion will answer that question. It will show why the first explanation was not the final explanation, and how this understanding prepares us for the next step in the series: the story of science making God unnecessary.

Sources

Boyer, Pascal. Religion Explained: The Evolutionary Origins of Religious Thought. Basic Books, 2001.

Dennett, Daniel C. Breaking the Spell: Religion as a Natural Phenomenon. Viking, 2006.

Feuerbach, Ludwig. The Essence of Christianity. 1841.

Hume, David. The Natural History of Religion. 1757.

Sagan, Carl. The Demon-Haunted World: Science as a Candle in the Dark. Random House, 1995.

Conclusion

The First Explanation Was Not the Final Explanation

Human beings created God because they were human.

They were afraid of death. They were frightened by nature. They were moved by dreams. They saw the dead in sleep and wondered if the dead still lived. They looked at thunder, drought, disease, childbirth, famine, earthquakes, eclipses, and the dark sky above them, and they searched for meaning before they had science.

They needed answers.

They needed comfort.

They needed order.

They needed hope.

They needed a story large enough to hold grief, fear, love, guilt, suffering, and the mystery of existence.

God gave them that story.

For a very long time, God explained what human beings could not yet explain. God explained the sky before astronomy. God explained disease before medicine. God explained death before neuroscience. God explained fertility before biology. God explained disaster before geology. God explained morality before secular ethics. God explained dreams before psychology. God explained suffering before human beings understood nature, chance, trauma, and social injustice.

This does not make early human beings foolish.

It makes them understandable.

They did not have microscopes, telescopes, laboratories, antibiotics, vaccines, satellites, weather science, evolutionary biology, genetics, or modern medicine. They had imagination, memory, fear, ritual, community, and story. With those tools, they created gods.

God was humanity's first great answer to the unknown.

But the first answer is not always the true answer.

A child may first think the moon follows him. A frightened person may first think thunder is anger. A patient may first think pain means certain death. A grieving person may first think a dream proves the dead are still speaking. First impressions feel powerful because they arrive before correction. They satisfy the mind before evidence has done its work.

Human history is, in many ways, the long process of replacing first impressions with better understanding.

This book has argued that God belongs to that history.

What This Book Has Shown

The idea of God did not arise from one single source. It grew from many human roots.

Fear of death created the dream of immortality.

Nature's violence created gods of sky, storm, river, fertility, and disaster.

Dreams created the feeling of another world.

The pattern-seeking brain saw intention behind events.

The social mind imagined invisible agents.

Ritual gave helplessness something to do.

Sacrifice turned fear into bargaining.

Tribe turned belief into belonging.

Authority turned God into power.

Stories turned mystery into memory.

Childhood turned belief into identity.

Emotion turned belief into certainty.

Together, these forces made God feel real.

This is why religion became so durable. It was never only an idea. It was a whole human system. It touched the body, the family, the dead, the tribe, the ruler, the child, the parent, the sick, the guilty, the poor, the dying, and the frightened.

Religion became powerful because it entered every major doorway of human life.

But when we examine these doorways carefully, we see something important. Each one can be explained naturally. We do not need to assume that God created religion. Human fear, grief, imagination, culture, authority, childhood, ritual, and social need explain religion very well.

Perhaps God did not create human religion.

Perhaps human religion created God.

That is the central argument of this book.

Understanding Religion Is Not the Same as Accepting It

Some people may think that criticizing religion means mocking believers.

That is not the purpose of this book.

A serious criticism of religion must begin with understanding. Believers are not simply ignorant. Many are kind, intelligent, loving, generous, and morally serious. Many have been shaped by family, culture, grief, fear, and hope. Many believe because belief was given to them before they could question it. Many hold faith because it is tied to parents, children, marriage, memory, community, and death.

To understand this is not weakness.

It is honesty.

If we do not understand why religion is powerful, we will not understand why it survives. If we do not understand the emotional needs behind belief, we will speak only to the surface. If we do not understand the role of family, childhood, ritual, music, and belonging, we will underestimate the depth of faith.

Religion is not a small error that can be removed by one clever argument. It is an ancient human structure built from fear, love, story, authority, hope, and identity.

That is exactly why it must be examined carefully.

The more powerful a belief is, the more honestly it should be questioned.

Comfort Cannot Decide Truth

Religion comforts because life is hard.

People die. Children suffer. Bodies fail. Injustice survives. Good people are hurt. Cruel people sometimes prosper. Love is interrupted by death. The future is uncertain. The universe is vast and silent. The human heart wants assurance.

Religion offers assurance.

It says the dead live on.

It says suffering has a purpose.

It says justice will come.

It says someone is watching.

It says someone cares.

It says life has a plan.

It says death is not the end.

These promises are emotionally powerful. They may help people endure tragedy. They may soften grief. They may give courage to the sick, the poor, the lonely, and the dying.

But comfort cannot decide truth.

A belief can comfort and still be false. A hope can be beautiful and still be unsupported. A story can help us survive and still not describe reality. A ritual can calm the body and still not change nature. A prayer can bring peace and still not prove that anyone is listening.

This is a difficult lesson, but it is necessary.

Reality does not become true because we need it to be true.

That statement may sound harsh at first, but it is also liberating. It frees us from being controlled by comforting illusions. It asks us to love truth more than certainty. It asks us to face life as it is, not only as we wish it to be.

This is the beginning of intellectual honesty.

A Meaningful Life Without God

If God was humanity's first explanation, some may fear that losing God means losing everything.

This fear is understandable, but it is not justified.

We do not lose the beauty of the sky when we understand astronomy.

We do not lose the wonder of life when we understand biology.

We do not lose love when we understand the brain.

We do not lose morality when we stop believing in divine punishment.

We do not lose grief when we stop believing in heaven.

We do not lose meaning when we stop believing in a cosmic plan.

Meaning is not destroyed when supernatural explanation is removed. Meaning becomes more human, more honest, and more urgent.

If this is the only life we know we have, then love matters more, not less.

If no divine hand will repair injustice, then human justice matters more, not less.

If no heaven will compensate every suffering child, then reducing suffering here matters more, not less.

If no God will save us from our own cruelty, then compassion, reason, and responsibility matter more, not less.

A godless life is not necessarily empty. It can be filled with love, truth, service, art, science, friendship, family, courage, kindness, and wonder.

The question is not whether life has meaning handed down from heaven.

The question is whether human beings can create and protect meaning here on Earth.

We can.

What We Gain by Letting Go of the First Explanation

Letting go of God as an explanation may feel like a loss. For some people, it is a painful loss. It may feel like losing a parent in the sky, a promise after death, a final court of justice, or a comforting story learned in childhood.

But we may also gain something important.

We gain honesty.

We gain freedom from fear-based belief.

We gain the courage to face death without illusion.

We gain responsibility for reducing suffering here, instead of explaining it away as divine will.

We gain the ability to build morality on compassion rather than command.

We gain the ability to see all human beings as members of one species, not as rival tribes favored or rejected by different gods.

We gain the right to question.

We gain the humility to admit what we do not know.

We gain the wonder of a universe that does not need to be supernatural in order to be magnificent.

The sky is still beautiful without a sky god.

Love is still meaningful without eternal reward.

Grief is still sacred without heaven.

Morality is still necessary without divine surveillance.

Wonder is still possible without worship.

Meaning is still real when human beings create it through love, truth, service, memory, and compassion.

To understand God as a human creation does not make life empty.

It makes life more honestly human.

From the First Explanation to Better Explanations

This book has shown why God was created.

The next question is how God became unnecessary.

That is the story of science.

Science did not defeat God by shouting. It did not remove God by anger. It did not make God unnecessary through slogans. It did something far more powerful.

It explained.

It explained the motion of planets.

It explained eclipses.

It explained disease.

It explained fossils.

It explained deep time.

It explained heredity.

It explained evolution.

It explained the brain.

It explained weather, earthquakes, infection, and countless other realities that were once placed under divine control.

Each explanation reduced the space where God was needed as an answer.

This did not happen in one moment. It happened slowly, painfully, and often against resistance. Copernicus moved

Earth from the center. Galileo looked into the sky and saw that the heavens were not perfect. Newton showed that the same laws governed Earth and sky. Hutton and Lyell revealed deep time. Darwin explained life without special creation. Pasteur and Koch explained disease without demons. Einstein and Hubble changed the universe. Modern neuroscience began to show how deeply mind depends on brain.

Again and again, the pattern repeated.

Where knowledge grew, God retreated.

That is not a slogan. It is a historical pattern.

The next book will follow that pattern in detail.

A Respectful Farewell to Humanity's First God

There is no need to hate the first explanation.

We can understand why it arose. We can understand why it comforted. We can understand why it survived. We can understand why people still hold it with deep emotion.

But understanding is not the same as believing.

Humanity once needed God to explain what it did not understand. Today, we no longer need that explanation in the same way. We have better tools. We have better methods. We have better evidence. We have science, history, psychology, medicine, anthropology, neuroscience, and philosophy.

These do not answer every question. They do not remove every sorrow. They do not make human life simple. But they are more honest than inventing invisible answers where evidence is missing.

God was humanity's first great explanation.

But the first explanation was not the final explanation.

Now the story must continue.

Closing Summary

This book has argued that God was created from human fear, death, nature, dreams, pattern-seeking, ritual, authority, culture, childhood, emotion, and story. These forces made religion psychologically natural and socially powerful. They explain why belief in God feels real to billions of people.

But none of these forces proves that God exists.

Human need explains belief.

It does not prove divine reality.

God was the answer human beings created before science gave them better answers. The next step is to examine how science gradually replaced God as an explanation.

That is where the next book begins.

Key Points

God was humanity's first great explanation for fear, death, nature, suffering, morality, and meaning.

Religion was psychologically natural because human beings needed comfort, control, hope, and belonging.

The God idea served many human needs, but usefulness does not prove truth.

Different cultures created different gods because religion was shaped by local human life.

Inherited belief often feels true because it enters early through family, ritual, fear, love, and repetition.

Religious feeling may be sincere, but sincerity does not prove external reality.

Understanding religion as a human creation does not make life empty.

Meaning, morality, love, and wonder can survive without God.

The next book will show how science made God unnecessary as an explanation.

Transition to Book 2

The first book has examined why humans created God.

The second book asks what happened when human beings began to understand nature more deeply.

The answer is one of the greatest stories in intellectual history.

As science advanced, God became less necessary.

The heavens no longer needed divine wheels. Disease no longer needed demons. Earth no longer needed sacred chronology. Life no longer needed special creation. Mind no longer needed soul as an explanation. Nature began to speak in laws, evidence, and causes.

Book 2 begins with that revolution.

Sources

Boyer, Pascal. Religion Explained: The Evolutionary Origins of Religious Thought. Basic Books, 2001.

Dennett, Daniel C. Breaking the Spell: Religion as a Natural Phenomenon. Viking, 2006.

Feuerbach, Ludwig. The Essence of Christianity. 1841.

Hume, David. The Natural History of Religion. 1757.

Sagan, Carl. The Demon-Haunted World: Science as a Candle in the Dark. Random House, 1995.

Appendix: Main Sources and Further Reading for Book 1

Why This Appendix Is Included

This book has argued that the idea of God did not appear in human life as a proven discovery. It grew from fear, death, dreams, nature, pattern-seeking, ritual, authority, culture, childhood, emotion, sacred story, and the human need for meaning.

This appendix is included for readers who want to examine these ideas more deeply. It is not meant to be a decorative list of books. It is meant to show that the argument of this book rests on a long tradition of serious inquiry.

The works listed here come from anthropology, psychology, sociology, philosophy, history, religious studies, myth studies, neuroscience, and science-based skepticism. Some authors are skeptical of religion. Some are sympathetic to religion. Some study religion mainly as a human, cultural, psychological, or historical phenomenon. Not every author listed here agrees with the conclusion of this book.

That is important.

A serious reader should not be afraid of disagreement. Good thinking grows when ideas are examined from more than one angle. The purpose of this appendix is to help the reader explore religion as a natural human phenomenon, not as a mystery that must be protected from questions.

Best Starting Points for General Readers

Readers who want a clear starting path may begin with the following books.

David Hume, The Natural History of Religion

Hume's short classic remains one of the best early attempts to explain religion through human fear, uncertainty, hope, and imagination. He helps us see religion not as a finished truth descending from heaven, but as a natural product of human life in a dangerous and poorly understood world.

Pascal Boyer, Religion Explained

Boyer's book is one of the most important modern works in the cognitive science of religion. He explains why the human mind is naturally prepared to notice agency, remember unusual beings, and transmit ideas about spirits, gods, and invisible powers.

Scott Atran, In Gods We Trust

Atran examines religion through anthropology, cognitive science, and evolution. His work is especially useful for understanding why religious belief can be emotionally powerful, socially durable, and difficult to dislodge.

Daniel Dennett, Breaking the Spell

Dennett argues that religion should be studied as a natural human phenomenon. His book is valuable because it treats religion as something that can be investigated, not something that must remain beyond inquiry.

Carl Sagan, The Demon-Haunted World

Sagan's book is one of the clearest defenses of science, skepticism, and critical thinking. It is essential reading for anyone who wants to understand why evidence matters and why human beings are so easily drawn to unsupported claims.

Michael Shermer, The Believing Brain

Shermer explains how people often form beliefs first and then search for evidence to support them. This is highly relevant to religion, especially in relation to pattern-seeking, confirmation bias, coincidence, and emotional certainty.

Classic Thinkers on Religion as a Human Creation

David Hume, The Natural History of Religion

Hume argued that religion grows from fear, hope, uncertainty, and attempts to explain nature. His work is central to the idea that gods arise from human need rather than clear divine revelation.

Ludwig Feuerbach, The Essence of Christianity

Feuerbach's central insight is one of the most important ideas behind this book: human beings often project their own needs, ideals, fears, and longings onto God. In simple terms, theology often reveals anthropology.

Sigmund Freud, The Future of an Illusion

Freud interpreted religion as a psychological response to helplessness, fear, and the longing for a father-like protector. Some of his ideas remain debated, but his view of religion as emotional protection remains influential.

Émile Durkheim, The Elementary Forms of Religious Life

Durkheim emphasized that religion is not only private belief. It is also social life, ritual, group identity, belonging, and collective meaning. His work helps explain why religion becomes community, not merely doctrine.

Max Weber, The Sociology of Religion

Weber studied the relationship between religion, authority, ethics, economics, and social organization. His work is useful for understanding how religion becomes connected with power, law, discipline, and institutions.

Karl Marx, selected writings on religion

Marx viewed religion partly as a response to suffering and social oppression. His work is useful because it shows how religion can comfort suffering people while sometimes protecting unjust social conditions.

Anthropology and the Early Study of Religion

Edward B. Tylor, Primitive Culture

Tylor helped develop the theory that early religion involved animism, spirits, souls, and invisible agents. Some of his language is dated today, but his work remains historically important.

James George Frazer, The Golden Bough

Frazer studied myth, magic, ritual, sacrifice, fertility, sacred kingship, and religious symbolism across cultures. Later scholars have criticized parts of his method, but the book remains a landmark in comparative religion.

Mircea Eliade, The Sacred and the Profane

Eliade explored how human beings divide the world into sacred and ordinary spaces. His work is helpful for understanding sacred places, ritual time, symbols, and the emotional power of religious experience.

Mircea Eliade, Myth and Reality

This book examines why myths matter to human beings and how sacred stories give structure to life, death, suffering, and identity.

Robert N. Bellah, Religion in Human Evolution

Bellah's major work explores religion across human evolution, symbolism, ritual, social development, and culture. It is a demanding book, but it is valuable for readers who want a deeper academic understanding.

Anthony F. C. Wallace, Religion: An Anthropological View

Wallace offers a broad anthropological view of religious belief, ritual, supernatural beings, healing, and religious specialists.

Cognitive Science and Psychology of Religion

Pascal Boyer, Religion Explained

Boyer explains how ordinary human mental habits help create and preserve belief in supernatural agents. This book is central to understanding why gods are easy to imagine, remember, and transmit.

Scott Atran, In Gods We Trust

Atran explores the relationship between religion, human cognition, emotion, evolution, and culture. His work is especially useful for understanding why religious beliefs are persistent across human societies.

Justin L. Barrett, Why Would Anyone Believe in God?

Barrett explains why the human mind may be naturally receptive to belief in gods and invisible agents. His conclusions are more sympathetic to religion than the position taken in this book, but his work remains useful.

Justin L. Barrett, Born Believers

This book focuses on children and religious belief. It is especially relevant to childhood, family, trust, and inherited belief.

Bruce Hood, SuperSense

Hood examines why human beings believe in invisible forces, hidden essences, souls, spirits, and unseen meanings. His work helps explain why supernatural thinking feels intuitive.

Stewart Elliott Guthrie, Faces in the Clouds

Guthrie's work is important for understanding anthropomorphism: the human tendency to see human-like agency in nature. This is directly relevant to the idea that people gave the universe a human face.

Daniel Kahneman, Thinking, Fast and Slow

Kahneman's work is not specifically about religion, but it is essential for understanding fast thinking, intuition, bias, and errors in judgment. It helps explain why feeling certain is not the same as being correct.

Leon Festinger, A Theory of Cognitive Dissonance

Festinger's work helps explain why people protect beliefs even when evidence challenges them. This is important for understanding religious certainty, defended doctrines, and failed prophecies.

Death, Immortality, and the Fear of Nonexistence

Ernest Becker, The Denial of Death

Becker's book is one of the most important works on the human fear of death. It helps explain why human beings create symbolic systems, heroic meanings, and dreams of immortality.

Douglas J. Davies, A Brief History of Death

Davies examines how human societies have understood death, mourning, funerals, and afterlife beliefs.

Robert Jay Lifton, The Broken Connection

Lifton explores the human need for symbolic continuity after death. His work is useful for understanding why people long for some form of survival beyond the body.

Elisabeth Kübler-Ross, On Death and Dying

This book is important for understanding the emotional and psychological experience of dying, grief, and end-of-life fear. It is not mainly an atheist book, but it remains historically influential.

Sheldon Solomon, Jeff Greenberg, and Tom Pyszczynski, The Worm at the Core

This book presents terror management theory in accessible language. It explains how awareness of death shapes culture, identity, belief, and human behavior.

Ritual, Community, and Social Identity

Victor Turner, The Ritual Process

Turner studied ritual, transition, symbolic action, and community. His work helps explain why rituals feel powerful during birth, marriage, illness, death, and social change.

Harvey Whitehouse, Modes of Religiosity

Whitehouse studies how religious rituals are remembered, repeated, and transmitted across communities.

Harvey Whitehouse, The Ritual Animal

This book explores ritual, imitation, group bonding, and social cohesion. It is useful for understanding how repeated religious acts join people together.

Ara Norenzayan, Big Gods

Norenzayan explores how belief in watchful, morally concerned gods may have helped large-scale human cooperation. His work is especially relevant to religion as moral surveillance and social control.

Jonathan Haidt, The Righteous Mind

Haidt examines moral psychology, loyalty, purity, authority, fairness, and group identity. His work helps explain why religion often becomes tied to moral emotion and tribal belonging.

Christian Smith, Religion: What It Is, How It Works, and Why It Matters

Smith provides a broad sociological view of religion as a living social and cultural system.

Myth, Sacred Story, and Religious Imagination

Joseph Campbell, The Hero with a Thousand Faces

Campbell studied recurring patterns in myths and hero stories across cultures. His work is useful for understanding why certain sacred story forms appear again and again.

Karen Armstrong, A Short History of Myth

Armstrong gives a sympathetic explanation of myth as a human search for meaning. Readers should compare her approach with the more skeptical approach of this book.

David A. Leeming, Creation Myths of the World

This work is helpful for readers who want examples of how different cultures explained the origin of the world.

Alan Dundes, Sacred Narrative

This collection examines myth as sacred story. It is helpful for understanding how communities preserve, repeat, and protect religious narratives.

Claude Lévi-Strauss, Myth and Meaning

Lévi-Strauss offers a structural approach to myth. His work helps readers understand how myths organize human thought.

Bruce Lincoln, Theorizing Myth

Lincoln examines myth, ideology, authority, and power. His work is useful for understanding how sacred stories can support political and social systems.

History of God, Scripture, and Religious Development

Karen Armstrong, A History of God

Armstrong's book is a widely read history of how ideas of God developed in Judaism, Christianity, and Islam. She is more sympathetic to religion than this book, but her historical overview is useful.

Ninian Smart, The World's Religions

Smart provides a broad overview of major religious traditions. This book helps readers understand religious diversity.

Huston Smith, The World's Religions

Smith gives a sympathetic introduction to world religions. It is useful for understanding how religions present themselves from within.

Elaine Pagels, The Gnostic Gospels

Pagels explores early Christian diversity and alternative religious texts. Her work is useful for understanding how religious traditions develop, compete, and become institutionalized.

Bart D. Ehrman, Misquoting Jesus

Ehrman explains textual transmission, copying, editing, and the human history of scripture.

Bart D. Ehrman, Jesus Interrupted

This book introduces readers to historical-critical study of the Bible and shows the difference between devotional reading and scholarly analysis.

Skepticism, Science, and Critical Thinking

Carl Sagan, The Demon-Haunted World

This is one of the most important modern books defending science, skepticism, and critical thinking. It is highly relevant to the spirit of the No God Required series.

Carl Sagan, Cosmos

Sagan presents the scientific view of the universe with beauty, humility, and wonder. He shows that science can inspire awe without requiring supernatural belief.

Daniel Dennett, Breaking the Spell

Dennett studies religion as a natural human phenomenon. His book is one of the most important modern skeptical works on religion.

Bertrand Russell, Why I Am Not a Christian

Russell's classic essay remains one of the clearest philosophical statements of religious skepticism.

Richard Dawkins, The God Delusion

Dawkins presents a strong atheist critique of God-belief and religious claims. His tone is sharper than the tone of this book, but his arguments are influential.

Christopher Hitchens, God Is Not Great

Hitchens offers a forceful literary critique of religion's moral, social, and political harms.

Sam Harris, The End of Faith

Harris argues that faith can be dangerous when it protects beliefs from evidence and criticism.

Victor Stenger, God: The Failed Hypothesis

Stenger argues that many God claims can be tested against scientific evidence and that they fail. This book will be especially useful in later books in the No God Required series.

Source Map by Theme

For readers who want to connect the major themes of this book with sources, the following map may help.

For fear, death, and immortality, begin with **Ernest Becker**, **Robert Jay Lifton**, **Douglas J. Davies**, and **The Worm at the Core**.

For dreams, spirits, souls, and animism, begin with **Edward B. Tylor**, **Pascal Boyer**, **Scott Atran**, **Bruce Hood**, and **Stewart Elliott Guthrie**.

For pattern-seeking, bias, and belief formation, begin with **Daniel Kahneman**, **Michael Shermer**, **Leon Festinger**, and **Daniel Dennett**.

For ritual, community, and belonging, begin with **Émile Durkheim**, **Victor Turner**, **Harvey Whitehouse**, **Jonathan Haidt**, and **Ara Norenzayan**.

For myth, sacred story, and religious imagination, begin with **Joseph Campbell**, **Karen Armstrong**, **David A. Leeming**, **Mircea Eliade**, and **Bruce Lincoln**.

For religious authority, institutions, and power, begin with **Max Weber**, **Émile Durkheim**, **Elaine Pagels**, and **Bart D. Ehrman**.

For scientific skepticism and critical thinking, begin with **Carl Sagan**, **Daniel Dennett**, **Bertrand Russell**, **Richard Dawkins**, **Sam Harris**, **Christopher Hitchens**, and **Victor Stenger**.

Suggested Reading Path

Readers who are new to this subject may follow this order.

First, read **David Hume's The Natural History of Religion**. It is short, classic, and directly relevant to the natural origin of religion.

Second, read **Pascal Boyer's Religion Explained** or **Scott Atran's In Gods We Trust**. These books introduce the cognitive and anthropological roots of religion.

Third, read **Ernest Becker's The Denial of Death** or **The Worm at the Core** to understand how fear of death shapes belief.

Fourth, read **Émile Durkheim's The Elementary Forms of Religious Life** or **Ara Norenzayan's Big Gods** to understand religion as social order and group identity.

Fifth, read **Michael Shermer's The Believing Brain** to understand pattern-seeking, confirmation bias, and belief formation.

Sixth, read **Carl Sagan's The Demon-Haunted World** for a beautiful and powerful defense of scientific thinking.

Finally, read **Daniel Dennett's Breaking the Spell** to understand religion as a natural phenomenon that can be studied, questioned, and explained.

Reliable Online and Academic Resources

Stanford Encyclopedia of Philosophy

A high-quality academic resource for philosophy of religion, atheism, agnosticism, divine hiddenness, the problem of evil, religious experience, and related topics.

Internet Encyclopedia of Philosophy

A useful resource for clear philosophical overviews written for students and general readers.

Pew Research Center Religion and Public Life

A reliable source for data on religious affiliation, secularization, global religious trends, and the rise of the religiously unaffiliated.

National Academies Resources

Useful for scientific material on evolution, science education, and the nature of scientific explanation.

NASA Science Resources

Useful for astronomy, cosmology, planets, stars, and the universe. These resources will become especially important in later books in this series.

NIH, PubMed, and NCBI

Reliable sources for medical, psychological, neurological, and biological research. These will be especially useful in later books on brain, mind, disease, and religious experience.

Final Note to the Reader

This appendix is not meant to end the discussion. It is meant to open the door.

The argument of this book is that religion can be understood as a human creation. That argument becomes stronger when we study anthropology, psychology, history, death, ritual, myth, childhood belief, social identity, and critical thinking together.

A serious reader should not fear difficult books or opposing viewpoints. Truth does not need protection from honest inquiry. If an idea is strong, examination will strengthen it. If an idea is weak, examination will reveal its weakness.

The next book in this series will move from the human origin of God to the scientific replacement of God.

It will ask how the great discoveries of science gradually made divine explanation unnecessary.

Thank You & Please Review This Book

Thank you for taking the time to read this book. I sincerely hope it has encouraged you to think more deeply, question more honestly, and look at religion, science, and human belief with greater clarity.

My goal in writing the **No God Required Series** is simple: to present serious, science-based, and easy-to-understand ideas about God, religion, nature, morality, and human meaning. These books are written for readers who value evidence, reason, intellectual honesty, and the courage to ask difficult questions.

If you found this book thoughtful, helpful, or worth reading, I would be truly grateful if you could leave a brief and honest review. Even a few sincere lines can make a meaningful difference. Your review helps other readers discover this series and encourages more open, rational, and evidence-based discussion about some of life's deepest questions.

Thank you again for your time, your trust, and your support.

Warm regards,
Dr. Prabhat Das

www.ingramcontent.com/pod-product-compliance
Lightning Source LLC
LaVergne TN
LVHW011704200926
848484LV00004B/1065
* 9 7 9 8 9 5 1 0 6 7 0 0 5 *